Mechanisms of flat weaving technology

The Textile Institute and Woodhead Publishing

The Textile Institute is a unique organization in textiles, clothing and footwear. Incorporated in England by a Royal Charter granted in 1925, the Institute has individual and corporate members in over 90 countries. The aim of the Institute is to facilitate learning, recognize achievement, reward excellence and disseminate information within the global textiles, clothing and footwear industries.

Historically, the Textile Institute has published books of interest to its members and the textile industry. To maintain this policy, the Institute has entered into partnership with Woodhead Publishing Limited to ensure that Institute members and the textile industry continue to have access to high calibre titles on textile science and technology.

Most Woodhead titles on textiles are now published in collaboration with the Textile Institute. Through this arrangement, the Institute provides an Editorial Board which advises Woodhead on appropriate titles for future publication and suggests possible editors and authors for these books. Each book published under this arrangement carries the Institute's logo.

Woodhead books published in collaboration with the Textile Institute are offered to Textile Institute members at a substantial discount. These books, together with those published by the Textile Institute that are still in print, are offered on the Woodhead website at: www.woodheadpublishing.com. Textile Institute books still in print are also available directly from the Institute's website at: www.textileinstitutebooks.com.

A list of Woodhead books on textile science and technology, most of which have been published in collaboration with the Textile Institute, can be found towards the end of the contents pages.

We are always happy to receive suggestions for new books from potential editors. To enquire about contributing to our Textiles series, please send your name, contact address and details of the topic/s you are interested in to sarah.lynch@woodheadpublishing.com. We look forward to hearing from you.

The team responsible for publishing this book:

Commissioning Editor: Kathryn Picking
Project Editor: Francis Dodds
Editorial and Production Manager: Mary Campbell
Production Editor: Adam Hooper
Project Manager: Newgen Knowledge Works Pvt Ltd
Copyeditor: Newgen Knowledge Works Pvt Ltd
Proofreader: Newgen Knowledge Works Pvt Ltd
Cover Designer: Terry Callanan

© Woodhead Publishing Limited, 2013

Woodhead Publishing Series in Textiles: Number 144

Mechanisms of flat weaving technology

Valeriy V. Choogin, Palitha Bandara
and Elena V. Chepelyuk

The Textile Institute

WOODHEAD
PUBLISHING

Oxford   Cambridge   Philadelphia   New Delhi

© Woodhead Publishing Limited, 2013

Published by Woodhead Publishing Limited in association with the Textile Institute
Woodhead Publishing Limited, 80 High Street, Sawston, Cambridge CB22 3HJ, UK
www.woodheadpublishing.com
www.woodheadpublishingonline.com

Woodhead Publishing, 1518 Walnut Street, Suite 1100, Philadelphia,
PA 19102–3406, USA

Woodhead Publishing India Private Limited, 303 Vardaan House, 7/28 Ansari Road,
Daryaganj, New Delhi – 110002, India
www.woodheadpublishingindia.com

First published 2013, Woodhead Publishing Limited
© Woodhead Publishing Limited, 2013. The publishers have made every effort to ensure that permission for copyright material has been obtained by authors wishing to use such material. The authors and the publishers will be glad to hear from any copyright holder it has not been possible to contact.
The authors have asserted their moral rights.

This book contains information obtained from authentic and highly-regarded sources. Reprinted material is quoted with permission, and sources are indicated. Reasonable efforts have been made to publish reliable data and information, but the authors and the publishers cannot assume responsibility for the validity of all materials. Neither the authors nor the publishers, nor anyone else associated with this publication, shall be liable for any loss, damage or liability directly or indirectly caused or alleged to be caused by this book.

Neither this book nor any part may be reproduced or transmitted in any form or by any means, electronic or mechanical, including photocopying, microfilming and recording, or by any information storage or retrieval system, without permission in writing from Woodhead Publishing Limited.

The consent of Woodhead Publishing Limited does not extend to copying for general distribution, for promotion, for creating new works, or for resale. Specific permission must be obtained in writing from Woodhead Publishing Limited for such copying.

Trademark notice: Product or corporate names may be trademarks or registered trademarks, and are used only for identification and explanation, without intent to infringe.

British Library Cataloguing in Publication Data
A catalogue record for this book is available from the British Library.

Library of Congress Control Number: 2013939357

ISBN 978-0-85709-780-4 (print)
ISBN 978-0-85709-785-9 (online)
ISSN 2042–0803 Woodhead Publishing Series in Textiles (print)
ISSN 2042–0811 Woodhead Publishing Series in Textiles (online)

The publisher's policy is to use permanent paper from mills that operate a sustainable forestry policy, and which has been manufactured from pulp which is processed using acid-free and elemental chlorine-free practices. Furthermore, the publisher ensures that the text paper and cover board used have met acceptable environmental accreditation standards.

Typeset by Newgen Knowledge Works Pvt Ltd, Chennai, India
Printed by Lightning Source

Contents

	Woodhead Publishing Series in Textiles	ix
	Preface	xv
	Note for students using this book	xix
1	**Introduction: classification and mechanisms of weaving machines**	**1**
1.1	Introduction	1
1.2	Classification of weaving machines	2
1.3	Basic mechanisms of the weaving machine	4
1.4	Elastic system of fabric formation (ESFF)	5
1.5	Advantages and disadvantages of different weaving machines	15
1.6	Questions for self-assessment	15
1.7	References	16
2	**Mechanisms of the weaving machine for warp release and warp tension control**	**17**
2.1	Introduction: mechanisms of the weaving machine for warp release and warp tension control	17
2.2	Warp brakes	20
2.3	Warp regulators	24
2.4	Condition of the equilibrium of the mechanism of a moving back-rest	34
2.5	Stabilization of the mode of release and the tensioning of warp threads	35
2.6	Comparative analysis of brakes and regulators	36
2.7	Questions for self-assessment	37
2.8	References	39
3	**Warp shedding in weaving: parameters and devices**	**40**
3.1	Introduction: parameters of the shed	40
3.2	Elongation of warp threads in shedding	44

3.3	The classification of shedding devices	49
3.4	Tappet shedding	49
3.5	Dobbies	54
3.6	The Jacquard machine	58
3.7	Comparative analysis of shedding devices	61
3.8	Questions for self-assessment	62
3.9	References	63
4	**The supply of weft on the weaving machine**	**64**
4.1	Introduction: the supply of weft on the weaving machine	64
4.2	Gauges for detecting the presence of weft on shuttle weaving machines	65
4.3	Battery type weft supply	67
4.4	Mechanisms for changing pirns	69
4.5	Safety devices of automatic pirn change	71
4.6	Change of hollow pirns	71
4.7	Change of shuttles	72
4.8	Multishuttle mechanisms	72
4.9	Weft supply on shuttleless weaving machines	76
4.10	Devices for measuring and control of weft tension	78
4.11	Questions for self-assessment	81
4.12	References	82
5	**Weft insertion**	**83**
5.1	Introduction: methods of weft insertion	83
5.2	Continuous insertion of weft by shuttle	88
5.3	Weft insertion by microshuttle	89
5.4	Weft insertion by projectile	92
5.5	Weft insertion by rapiers	95
5.6	Weft insertion by air and water jets	100
5.7	Pneumatic-rapier weft insertion	103
5.8	Weft insertion by an electromagnetic drive	106
5.9	Weft insertion by the inertial method	106
5.10	Comparative analysis of different methods of weft insertion	106
5.11	Questions for self-assessment	107
5.12	References	108
6	**Woven fabric formation: principles and methods**	**109**
6.1	Introduction: woven fabric formation	109
6.2	Fabric-forming mechanisms	112
6.3	Formation of the woven fabric cell	114
6.4	Parameters of woven fabric formation	116

6.5	Ring temples	121
6.6	Methods of easing of fabric formation	122
6.7	Comparative analysis of the methods of fabric forming	123
6.8	Questions for self-assessment	123
6.9	References	124
7	**Mechanisms for woven fabric take-up**	**125**
7.1	Introduction: mechanisms for woven fabric take-up	125
7.2	Winding woven fabric on the cloth beam	131
7.3	Comparative analysis of methods of woven fabric take-up	132
7.4	Questions for self-assessment	133
7.5	References	134
8	**Safety devices on weaving machines**	**135**
8.1	Introduction: safety devices on weaving machines	135
8.2	Weft controllers	137
8.3	Devices for prevention of warp thread breakage	142
8.4	Comparative analysis of safety devices	144
8.5	Questions for self-assessment	144
8.6	References	144
9	**Weaving machine drives: mechanisms and types**	**145**
9.1	Introduction: the weaving machine drive	145
9.2	Mechanisms for driving the main shaft of a weaving machine	146
9.3	Weaving machine brakes	148
9.4	Combined start-up and braking mechanisms	149
9.5	Comparative analysis of different loom drives	149
9.6	Questions for self-assessment	150
9.7	References	150
10	**Weaving machine parameters for specific woven fabric structures**	**151**
10.1	Introduction: the normalization process for weaving operations	151
10.2	Estimating drawing-in parameters for a weaving machine	153
10.3	Setting up parameters for the machine–yarn–fabric path (MYFP) on weaving machines	155
10.4	Verification of parameters for the elastic system of fabric formation (ESFF) and machine–yarn–fabric path (MYFP) on weaving machines	156
10.5	Evaluating warp thread tension by oscillogram analysis	157

10.6	Coordination of weaving cycles	165
10.7	Factors affecting the productivity of weaving machines	166
10.8	Comparing operating conditions for natural and synthetic fibres	168
10.9	Questions for self-assessment	169
10.10	References	170
11	**Control of woven fabric quality: defects and quality assurance of grey fabrics**	**171**
11.1	Introduction: the quality control of woven fabric	171
11.2	Defects in grey fabric	172
11.3	Quality assurance of grey fabric	174
11.4	Equipment for the control of woven fabric quality	175
11.5	Questions for self-assessment	177
11.6	References	178
12	**Movement of raw materials and finished fabrics in weaving manufacture**	**179**
12.1	Introduction: product transportation in weaving manufacture	179
12.2	Transportation of raw materials and outputs	179
12.3	Questions for self-assessment	181
12.4	References	181
	Appendix 1: Further reading on weaving technology	*183*
	Appendix 2: Glossary of terms applied to weaving machines and weaving technology	*185*
	Index	*205*

Woodhead Publishing Series in Textiles

1 **Watson's textile design and colour (Seventh edition)**
  Edited by Z. Grosicki
2 **Watson's advanced textile design**
  Edited by Z. Grosicki
3 **Weaving (Second edition)**
  P. R. Lord and M. H. Mohamed
4 **Handbook of textile fibres. Volume 1: Natural fibres**
  J. Gordon Cook
5 **Handbook of textile fibres. Volume 2: Man-made fibres**
  J. Gordon Cook
6 **Recycling textile and plastic waste**
  Edited by A. R. Horrocks
7 **New fibers (Second edition)**
  T. Hongu and G. O. Phillips
8 **Atlas of fibre fracture and damage to textiles (Second edition)**
  J. W. S. Hearle, B. Lomas and W. D. Cooke
9 **Ecotextile '98**
  Edited by A. R. Horrocks
10 **Physical testing of textiles**
  B. P. Saville
11 **Geometric symmetry in patterns and tilings**
  C. E. Horne
12 **Handbook of technical textiles**
  Edited by A. R. Horrocks and S. C. Anand
13 **Textiles in automotive engineering**
  W. Fung and J. M. Hardcastle
14 **Handbook of textile design**
  J. Wilson
15 **High-performance fibres**
  Edited by J. W. S. Hearle
16 **Knitting technology (Third edition)**
  D. J. Spencer
17 **Medical textiles**
  Edited by S. C. Anand
18 **Regenerated cellulose fibres**
  Edited by C. Woodings
19 **Silk, mohair, cashmere and other luxury fibres**
  Edited by R. R. Franck
20 **Smart fibres, fabrics and clothing**
  Edited by X. M. Tao
21 **Yarn texturing technology**
  J. W. S. Hearle, L. Hollick and D. K. Wilson
22 **Encyclopedia of textile finishing**
  H-K. Rouette
23 **Coated and laminated textiles**
  W. Fung

24 **Fancy yarns**
   R. H. Gong and R. M. Wright
25 **Wool: Science and technology**
   Edited by W. S. Simpson and G. Crawshaw
26 **Dictionary of textile finishing**
   H-K. Rouette
27 **Environmental impact of textiles**
   K. Slater
28 **Handbook of yarn production**
   P. R. Lord
29 **Textile processing with enzymes**
   Edited by A. Cavaco-Paulo and G. Gübitz
30 **The China and Hong Kong denim industry**
   Y. Li, L. Yao and K. W. Yeung
31 **The World Trade Organization and international denim trading**
   Y. Li, Y. Shen, L. Yao and E. Newton
32 **Chemical finishing of textiles**
   W. D. Schindler and P. J. Hauser
33 **Clothing appearance and fit**
   J. Fan, W. Yu and L. Hunter
34 **Handbook of fibre rope technology**
   H. A. McKenna, J. W. S. Hearle and N. O'Hear
35 **Structure and mechanics of woven fabrics**
   J. Hu
36 **Synthetic fibres: Nylon, polyester, acrylic, polyolefin**
   Edited by J. E. McIntyre
37 **Woollen and worsted woven fabric design**
   E. G. Gilligan
38 **Analytical electrochemistry in textiles**
   P. Westbroek, G. Priniotakis and P. Kiekens
39 **Bast and other plant fibres**
   R. R. Franck
40 **Chemical testing of textiles**
   Edited by Q. Fan
41 **Design and manufacture of textile composites**
   Edited by A. C. Long
42 **Effect of mechanical and physical properties on fabric hand**
   Edited by H. M. Behery
43 **New millennium fibers**
   T. Hongu, M. Takigami and G. O. Phillips
44 **Textiles for protection**
   Edited by R. A. Scott
45 **Textiles in sport**
   Edited by R. Shishoo
46 **Wearable electronics and photonics**
   Edited by X. M. Tao
47 **Biodegradable and sustainable fibres**
   Edited by R. S. Blackburn
48 **Medical textiles and biomaterials for healthcare**
   Edited by S. C. Anand, M. Miraftab, S. Rajendran and J. F. Kennedy
49 **Total colour management in textiles**
   Edited by J. Xin
50 **Recycling in textiles**
   Edited by Y. Wang
51 **Clothing biosensory engineering**
   Y. Li and A. S. W. Wong
52 **Biomechanical engineering of textiles and clothing**
   Edited by Y. Li and D. X-Q. Dai
53 **Digital printing of textiles**
   Edited by H. Ujiie

54 **Intelligent textiles and clothing**
   Edited by H. R. Mattila
55 **Innovation and technology of women's intimate apparel**
   W. Yu, J. Fan, S. C. Harlock and S. P. Ng
56 **Thermal and moisture transport in fibrous materials**
   Edited by N. Pan and P. Gibson
57 **Geosynthetics in civil engineering**
   Edited by R. W. Sarsby
58 **Handbook of nonwovens**
   Edited by S. Russell
59 **Cotton: Science and technology**
   Edited by S. Gordon and Y.-L. Hsieh
60 **Ecotextiles**
   Edited by M. Miraftab and A. R. Horrocks
61 **Composite forming technologies**
   Edited by A. C. Long
62 **Plasma technology for textiles**
   Edited by R. Shishoo
63 **Smart textiles for medicine and healthcare**
   Edited by L. Van Langenhove
64 **Sizing in clothing**
   Edited by S. Ashdown
65 **Shape memory polymers and textiles**
   J. Hu
66 **Environmental aspects of textile dyeing**
   Edited by R. Christie
67 **Nanofibers and nanotechnology in textiles**
   Edited by P. Brown and K. Stevens
68 **Physical properties of textile fibres (Fourth edition)**
   W. E. Morton and J. W. S. Hearle
69 **Advances in apparel production**
   Edited by C. Fairhurst
70 **Advances in fire retardant materials**
   Edited by A. R. Horrocks and D. Price
71 **Polyesters and polyamides**
   Edited by B. L. Deopura, R. Alagirusamy, M. Joshi and B. S. Gupta
72 **Advances in wool technology**
   Edited by N. A. G. Johnson and I. Russell
73 **Military textiles**
   Edited by E. Wilusz
74 **3D fibrous assemblies: Properties, applications and modelling of three-dimensional textile structures**
   J. Hu
75 **Medical and healthcare textiles**
   Edited by S. C. Anand, J. F. Kennedy, M. Miraftab and S. Rajendran
76 **Fabric testing**
   Edited by J. Hu
77 **Biologically inspired textiles**
   Edited by A. Abbott and M. Ellison
78 **Friction in textile materials**
   Edited by B. S. Gupta
79 **Textile advances in the automotive industry**
   Edited by R. Shishoo
80 **Structure and mechanics of textile fibre assemblies**
   Edited by P. Schwartz
81 **Engineering textiles: Integrating the design and manufacture of textile products**
   Edited by Y. E. El-Mogahzy
82 **Polyolefin fibres: Industrial and medical applications**
   Edited by S. C. O. Ugbolue

83 **Smart clothes and wearable technology**
   Edited by J. McCann and D. Bryson
84 **Identification of textile fibres**
   Edited by M. Houck
85 **Advanced textiles for wound care**
   Edited by S. Rajendran
86 **Fatigue failure of textile fibres**
   Edited by M. Miraftab
87 **Advances in carpet technology**
   Edited by K. Goswami
88 **Handbook of textile fibre structure. Volume 1 and Volume 2**
   Edited by S. J. Eichhorn, J. W. S. Hearle, M. Jaffe and T. Kikutani
89 **Advances in knitting technology**
   Edited by K-F. Au
90 **Smart textile coatings and laminates**
   Edited by W. C. Smith
91 **Handbook of tensile properties of textile and technical fibres**
   Edited by A. R. Bunsell
92 **Interior textiles: Design and developments**
   Edited by T. Rowe
93 **Textiles for cold weather apparel**
   Edited by J. T. Williams
94 **Modelling and predicting textile behaviour**
   Edited by X. Chen
95 **Textiles, polymers and composites for buildings**
   Edited by G. Pohl
96 **Engineering apparel fabrics and garments**
   J. Fan and L. Hunter
97 **Surface modification of textiles**
   Edited by Q. Wei
98 **Sustainable textiles**
   Edited by R. S. Blackburn
99 **Advances in yarn spinning technology**
   Edited by C. A. Lawrence
100 **Handbook of medical textiles**
   Edited by V. T. Bartels
101 **Technical textile yarns**
   Edited by R. Alagirusamy and A. Das
102 **Applications of nonwovens in technical textiles**
   Edited by R. A. Chapman
103 **Colour measurement: Principles, advances and industrial applications**
   Edited by M. L. Gulrajani
104 **Fibrous and composite materials for civil engineering applications**
   Edited by R. Fangueiro
105 **New product development in textiles: Innovation and production**
   Edited by L. Horne
106 **Improving comfort in clothing**
   Edited by G. Song
107 **Advances in textile biotechnology**
   Edited by V. A. Nierstrasz and A. Cavaco-Paulo
108 **Textiles for hygiene and infection control**
   Edited by B. McCarthy
109 **Nanofunctional textiles**
   Edited by Y. Li
110 **Joining textiles: Principles and applications**
   Edited by I. Jones and G. Stylios
111 **Soft computing in textile engineering**
   Edited by A. Majumdar

112 **Textile design**
   Edited by A. Briggs-Goode and K. Townsend
113 **Biotextiles as medical implants**
   Edited by M. W. King and B. S. Gupta and R. Guidon
114 **Textile thermal bioengineering**
   Edited by Y. Li
115 **Woven textile structure**
   B. K. Behera and P. K. Hari
116 **Handbook of textile and industrial dyeing. Volume 1: Principles, processes and types of dyes**
   Edited by M. Clark
117 **Handbook of textile and industrial dyeing. Volume 2: Applications of dyes**
   Edited by M. Clark
118 **Handbook of natural fibres. Volume 1: Types, properties and factors affecting breeding and cultivation**
   Edited by R. Kozłowski
119 **Handbook of natural fibres. Volume 2: Processing and applications**
   Edited by R. Kozłowski
120 **Functional textiles for improved performance, protection and health**
   Edited by N. Pan and G. Sun
121 **Computer technology for textiles and apparel**
   Edited by J. Hu
122 **Advances in military textiles and personal equipment**
   Edited by E. Sparks
123 **Specialist yarn and fabric structures**
   Edited by R. H. Gong
124 **Handbook of sustainable textile production**
   M. I. Tobler-Rohr
125 **Woven textiles: Principles, developments and applications**
   Edited by K. Gandhi
126 **Textiles and fashion: Materials design and technology**
   Edited by R. Sinclair
127 **Industrial cutting of textile materials**
   I. Viļumsone-Nemes
128 **Colour design: Theories and applications**
   Edited by J. Best
129 **False twist textured yarns**
   C. Atkinson
130 **Modelling, simulation and control of the dyeing process**
   R. Shamey and X. Zhao
131 **Process control in textile manufacturing**
   Edited by A. Majumdar, A. Das, R. Alagirusamy and V. K. Kothari
132 **Understanding and improving the durability of textiles**
   Edited by P. A. Annis
133 **Smart textiles for protection**
   Edited by R. A. Chapman
134 **Functional nanofibers and applications**
   Edited by Q. Wei
135 **The global textile and clothing industry: Technological advances and future challenges**
   Edited by R. Shishoo
136 **Simulation in textile technology: Theory and applications**
   Edited by D. Veit
137 **Pattern cutting for clothing using CAD: How to use Lectra Modaris pattern cutting software**
   M. Stott
138 **Advances in the dyeing and finishing of technical textiles**
   M. L. Gulrajani
139 **Multidisciplinary know-how for smart textiles developers**
   Edited by T. Kirstein

140 **Handbook of fire resistant textiles**
    Edited by F. Selcen Kilinc
141 **Handbook of footwear design and manufacture**
    Edited by A. Luximon
142 **Textile-led design for the active ageing population**
    Edited by J. McCann and D. Bryson
143 **Optimizing decision making in the apparel supply chain using artificial intelligence (AI): From production to retail**
    W. K. Wong, Z. X. Guo and S. Y. S. Leung
144 **Mechanisms of flat weaving technology**
    V. V. Choogin, P. Bandara and E. V. Chepelyuk
145 **Innovative jacquard textile design using digital technologies**
    F. Ng and J. Zhou
146 **Advances in shape memory polymers**
    J. Hu
147 **Design of clothing manufacturing processes: A systematic approach to planning, scheduling and control**
    J. Geršak
148 **Anthropometry, apparel sizing and design**
    D. Gupta and N. Zakaria
149 **Silk: Processing, properties and applications**
    K. Murugesh Babu
150 **Advances in filament spinning**
    D. Zhang
151 **Designing apparel for consumers: The impact of body shape and size**
    M.-E. Faust and S. Carrier

Preface

People cannot go without clothes, and many industries cannot produce goods without the use of woven materials. That is why the need for specialists skilled in weaving will always exist. To create an effective woven fabric structure, a designer must understand its specific purpose. A process engineer must then choose the appropriate type of weaving machine to produce the specific woven fabric structure and determine the particular parameters of the elastic system of fabric formation to be applied. It is also essential to know the purpose of all the mechanisms of the chosen type of weaving machine. Weaving machines consist, for example, of mechanisms of different types: levers, cams, frictional drives, cylindrical and worm reduction gears, differential gears, etc. A range of pneumatic and hydraulic devices, lever mechanisms and electromagnetic devices are used.

In addition to understanding weaving machines, the process engineer or student of weaving must understand the conditions of tension on the warp, the physical capacity of the warp and weft for bending, crimping, stretching, etc. To understand the conditions of warp and weft deformation and tensile behaviour, the filling parameters of the elastic system of fabric formation, engineers and students need a good grounding in mathematics, and a basic knowledge of mechanics and theories of the strength of materials. They can compare weaving technology with examples from general engineering. The study of the different types of reduction gears in weaving machines can be compared with the general mechanics of reduction gears; beam theory in the 'theory of the strength of materials' can be used to calculate thread bending in a woven fabric structure. Within 'higher mathematics', compounding differential and integral equations makes it possible to study the frictional movement of a thread along the cylindrical surface of the weaving machine's back-rest, etc.

Effective learning is best if it uses practical examples, and debate is encouraged. In 1903, for example, the outstanding engineer and teacher in the field of weaving, A.G. Lapisov, wrote in the Introduction to his book *Directions on the analysis of woven fabrics and the elaboration of the weaving plan* about giving the opportunity to students 'to discuss together, ... to exchange thoughts and conclusions concerning current and forthcoming technologies ...'. Such exchange of thoughts between students, with the assistance and participation of the teacher, leads to students acquiring a greater knowledge and deeper understanding in a shorter period compared to direct explanation by the teacher. It is also important for teachers to help students learn effectively by developing the ability for comparative *analysis* of any phenomena, whether technological processes, mechanisms, parameters and properties of products. The ability to analyze the strengths and weaknesses of one mechanism by comparing it to another, for example, leads to better understanding and a more rational solution to a problem.

This book presents essential information to allow effective process management in the field of weaving manufacture. Chapter 1 introduces the reader to the classification of different types of weaving machines, the basic mechanisms involved and the process of 2-D (two-dimensional, or flat) woven fabric formation. The methods of calculation of quantities are given that relate to the elastic system of 2-D fabric formation; such as the tension and elastic deformation of threads in the let-off zone of the warp beam, the effect of friction of warp yarns and woven fabric on a stationary cylindrical surface, and the definition of the equivalent thread and fabric length in the elastic system of fabric formation.

Chapter 2 discusses the principles and mechanisms of warp release (warp let-off). Warp beam brakes with manual or automatic regulation of the braking force are reviewed. The chapter describes how the warp is released in the required length into the weaving zone, rather than simply being unwound from the warp beam. It discusses the operation of the beam regulator, which handles equal and unequal thread tension coming from two warp beams on one weaving machine by means of the differential. At the end of the chapter, there is an example of the comparative analysis of different release mechanisms, as well as recommendations on how to use them.

Chapter 3 provides a description of the geometry of the three parts of the warp shed, extending from the fell of the woven fabric to the back-rest. On this basis, different types of shed, phases of their formation, and warp thread elongation caused by the movement of the healds are examined and ways of reducing elongation are identified. The basic types of shedding devices, dobby and Jacquard machines, are described.

Chapter 4 reviews how the supply of weft is maintained on a weaving machine. It describes methods and devices used for maintaining the weft supply on shuttle weaving machines and multishuttle mechanisms, and methods of weft supply on shuttleless weaving machines.

Chapter 5 describes the different methods of weft insertion used on weaving machines. These include the use of shuttle, rapier, projectile, air-jet and water-jet methods. Weft insertion based on electromagnetic, pneumatic-rapier, microshuttle, and weft inertia methods are also discussed.

Chapter 6 describes woven fabric formation on both shuttle and shuttleless weaving machines. It is important to understand the peculiarities of the conditions of woven fabric formation using the different mechanisms employed: the methods of moving the weft into the fell of the woven fabric; and methods of weft beat-up: front beating-up, and point beating-up. A method of calculation of parameters of woven fabric formation is given for using in practice. It is also important to understand the unique properties of the elastic system of fabric formation, particularly the automatic regulation of warp tension on the weaving machine. At the end of the chapter, methods of reducing the tension of the elastic system of fabric formation are covered.

The characteristics of fabric take-up from the working area and its winding onto the cloth beam are discussed in Chapter 7. This chapter reviews different types of take-up motion regulators which allow the formation of woven fabric with various arrangements of structure using weft with varied longitudinal evenness in thickness.

Chapter 8 describes safety (protective) devices provided on woven machines: warp stop motions, weft detectors and safety devices against the breakage of warp threads. The different types of driving and stopping mechanisms for shuttle and shuttleless weaving machines are presented in Chapter 9.

Chapter 10 discusses ways of estimating the optimal parameters of weaving machine settings with regard to specific woven fabric structures, the coordination of the operation of weaving machine mechanisms, and appraisal of weaving machine productivity in different units of measurement at different speeds of weaving machine operation as required.

Chapter 11 describes the methods and special equipment used for the evaluation of the quality and the quantity of the woven fabric produced. Chapter 12 describes the use of electrically-driven vehicles for the transportation of raw materials and outputs between different units of production at the weaving factory. Finally, a General Bibliography is presented in Appendix 1, while a Glossary of terms applied to weaving machines and weaving technology is presented in Appendix 2.

We are grateful to the colleagues who have helped us with this book: Professor Y. Bardachov, Rector of Kherson National Technical University (KNTU), and Mrs T. Gnedko, Lecturer at KNTU, for their valuable assistance rendered in the preparation of this work. We owe our thanks to numerous specialists in the field of weaving whose suggestions and advice have made this book possible. We are also grateful to the publishers for the consideration they have given to our wishes in the preparation of the book. We hope that this book will find readers not only among the students and teachers but also among specialists in weaving manufacture.

Valeriy V. Choogin *Palitha Bandara* *Elena V. Chepelyuk*
Kherson, Ukraine *Leeds, UK* *Kherson, Ukraine*

Note for students using this book

The target of learning is the ability to solve practical problems. The full benefit of studying is achieved only if the student participates in discussion of the topic, and tries to offer his or her own solutions to a problem. Student learning is enhanced by discussion of the advantages and disadvantages of different technologies and mechanisms. All active participants in a discussion will remember more and acquire the habits of analysis, problem-solving and decision-making. As a result, the level of knowledge of such students tends to be considerably higher than that of those who receive tuition by correspondence.

Independent work by the student consolidates the knowledge that has been gained in lessons. Such work should be done without delay with the help of training material from different authors. It is recommended that students look through this previous material and discover its connection with the topic being studied. It is useful then to write down the resulting ideas and questions. This will help students to formulate their ideas clearly and concisely.

This book does not claim to be an exhaustive reference book on all existing types of weaving machines and their constructions. Based on our experience, we have limited ourselves to reviewing the basic mechanisms, and the advantages and disadvantages of the main mechanisms in weaving machines. Clear illustrations are provided to help in the understanding of the construction and functioning of each mechanism.

We suggest students use this textbook in the following way:

- The student should assess the basic principles of each mechanism as initially developed. It is important to note that some later variants lack the flexibility to cope with certain novel or intricate fabric structures. Students should use the opportunity to design and offer new solutions for a mechanism to improve its functionality based on first principles.
- At the end of each chapter of the textbook, there is an example of a comparative analysis of mechanisms of a different construction together with recommendations for improvement. The student can add to this analysis and formulate his or her own recommendations.

- There are self-evaluation questions at the end of each chapter. The student's the answers can be checked by referring back to the text of the related chapter.

We hope that this book will help you to study a weaving course successfully and develop the qualities of a *thinker*. This critical ability will help you to become a successful professional in the field of weaving.

1
Introduction: classification and mechanisms of weaving machines

DOI: 10.1533/9780857097859.1

Abstract: This chapter provides general information on weaving machines. It introduces the reader to the different types of weaving machine, the basic mechanisms involved and the process of creating 2-D (two-dimensional, or flat) woven fabric. The next section discusses the calculation of quantities related to the elastic system of 2-D fabric manufacture, such as the tension and elastic deformation of threads in the let-off zone of the warp beam, the effect of friction on warp yarns and woven fabric on a stationary cylindrical surface, and the definition of the equivalent thread and fabric length.

Key words: different types of weaving machine, elastic system of fabric formation (ESFF).

1.1 Introduction

The basic weaving operation involves a warp; this consists of a set of evenly spaced parallel warp yarns into which individual transverse weft yarns are sequentially inserted, so as to obtain a suitably interlaced structure which is called a 'woven fabric'. The woven structure is determined by the order or pattern in which individual warp yarns are made to interlace with the weft yarns. Before each weft yarn is inserted, the individual warp yarns are raised or lowered as determined by the required fabric structure. The raising or lowering of the warp yarns, known as 'shedding', forms a diamond-shaped opening across the warp called the 'shed', through which the weft yarn is inserted. Each weft yarn inserted is pushed or 'beaten-up' against the previously inserted weft by means of a movable reed, which consists of a uniformly spaced set of plates or wires. By repeating these operations cyclically, a woven fabric with a distinct uniform structure is formed.

Weaving machines have been developed to enable the industrial production of woven fabrics in large volumes at high weaving speeds. This requires the preparation of warp and weft yarns of suitable quality, and various preparatory operations (winding, warping, sizing, drawing-in the warp yarns through the droppers, the healds of the harness and the reed). These operations are essential to ensure that high-quality fabric can be produced with minimum interruptions to production.

After carrying out the necessary preparation, the warp and weft threads are transferred to the weaving machines in the production unit. They are then interlaced to form a woven fabric having a particular density, width and type of interlacing, so as to be able to achieve particular physical characteristics (such as air permeability, abrasion resistance, tensile strength, crease retention, draping, etc.) as required for the fabric's intended application.

1.2 Classification of weaving machines

Weaving machines can be classified according to several characteristics:

- nature of fabric formation;
- method of weft insertion (picking) used;
- type of shedding device used;
- number of colours of the weft threads that can be used;
- type of yarn used;
- width of the woven fabric produced;
- number of sheds simultaneously formed, along the warp on the weaving machine.

Let us now discuss these classifications.

- **The nature of fabric formation:** Most weaving machines carry out **periodic formation** of fabric. Each period can be considered as starting with the shed opening operation, followed by the insertion of the weft across the full width of the warp. The closing of the shed produces the required interlacing of the warp with the newly-inserted weft. The beating-up operation then consolidates the new weft into the fabric. The element of new fabric thus formed is then drawn off, and a fresh warp released into the weaving zone. Weaving continues as this sequence of operations is repeated. There are also weaving machines which carry out **continuous formation** of fabric, where several weft threads are progressively inserted through a multiwave-shed, followed by an arrangement of 'point beat-up'.

    In the case of periodic formation, the shed is formed simultaneously across the full width of the warp. A single weft thread is then inserted across the full width of the warp and is followed by the beat-up operation. This process is repeated periodically at a suitably high rate. The majority of weaving machine types fall into this category. These are also known as 'single phase' weaving machines (e.g. Sulzer, Dornier, Picanol, Vamatex).

    In continuous formation, the shed is made in each of several groups or sections of equal width across the overall width of the warp. The opening and closing of the shed in successive groups is timed so that a number of picks equal to the number of groups of warp can be simultaneously conveyed across the warp. Thus, beating-up takes place progressively across the width of the warp, for each weft thread inserted. This category of weaving machine is described as 'multiphase' (e.g. Wasserman, Cerdans, Ruti, etc.).

- **Method of weft insertion (picking): Shuttle weaving machines** (e.g. Picanol) pick the weft thread into the shed by means of a shuttle which carries the weft package; **shuttleless weaving machines** use stationary thread packages positioned on one side of the warp, from which the weft is drawn through the shed by means of one of several different devices. So, shuttleless weaving machines can be classified according to the method of weft insertion (picking), as follows:
  - rapier weaving machines (Draper, Vamatex, Ruti, P-190, etc.);
  - projectile weaving machines (Sulzer, Novostav, etc.);
  - pneumatic (air-jet) weaving machines (Elitex, Sulzer, Nissan, etc.);
  - water-jet weaving machines (Elitex, Nissan, etc.);
  - pneumatic-rapier weaving machines (ATPR).
- **Type of the shedding device:** There are various shedding mechanisms available: cam, dobby or Jacquard.
- **The number of colours of weft threads that can be used:** There are single-colour (monoweft) and multicolour (multiweft) weaving machines.
- **Type of yarn:** Weaving machines are able to produce fabrics from a variety of yarns: cotton, wool, silk, flax, synthetics, glass, etc.
- **Width of the woven fabric produced:** Weaving machines can produce fabrics with working widths ranging from narrow (up to 120 cm) to wide (up to 5 m).
- **Number of sheds simultaneously formed along the warp on the weaving machine:** There are monoshed and multished weaving machines.
- Additionally, shuttle weaving machines are classified according to:
  - the method of weft replenishment: on automatic machines, the weft package (or shuttle, depending on the design of the machine) is replaced automatically in the event of weft breakage or depletion; and on non-automatic weaving machines, the weft package is changed manually;
  - the construction of the picking mechanism: either middle- or bottom-picking;
  - the number of shuttles used: monoshuttle or multishuttle.

The development of shuttleless weaving led to devising of other methods of picking the weft, for example, rapier picking, which is now widely used on modern weaving machines. Weaving machine manufacturers have generally identified their weaving machines by a combination of letters and numbers, but there is no common system with which to identify a machine's type.

1.3 Basic mechanisms of the weaving machine

Weaving machine manufacturers each have their own versions of the major working mechanisms, but weaving machines are similar in the manner in which the woven fabric is formed by the addition of successive picks into the cloth-fell, which is the boundary between the warp yarns and the freshly formed fabric.

Figure 1.1 shows the main components of a weaving machine: the warp beam (1), which carries the supply of warp threads; back-rest (2); droppers (3); harness (4); reed (5); temples (6); breast beam (7); emery roller (8); pressure roller (9); and cloth beam (10). Let-off (releasing) of the warp from the warp beam (1) is carried out by the regulator (11), and movement of the harness (the collection of heald shafts) (4) by the shedding device (12). The movement of the reed (5) is provided by the sley action (13). The take-up action (15) draws off the woven fabric (16) and the fabric is wound onto the cloth beam (10). The warp and the woven fabric pass over a number of guides: the back-rest (2), temples (6), breast rail (7) and pressure roller (9). Some of these components carry out additional functions. Auxiliary mechanisms are provided on a weaving machine to prevent defects in the fabric and to ensure safety of operation. The movement of all the weaving machine mechanisms is powered by an electric motor.

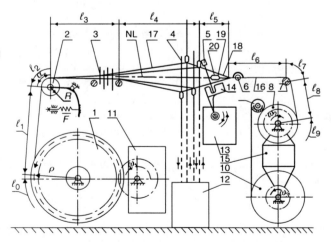

1.1 Machine–yarn–fabric path (MYFP) of the weaving machine. Refer to text for detailed explanation of components.

Introduction: classification and mechanisms of weaving machines 5

Fabric formation on a weaving machine consists of the following basic, cyclically repeating operations:

- warp threads (17) are raised or lowered vertically, according to the requirements of the fabric structure, to form a spatial quadrangle which is called the 'shed'; this operation is called 'shedding';
- the weft thread (20) is picked (inserted) into the open shed (18) and moved along the shed by the picking device (19);
- the inserted weft (20) is beaten-up to the cloth-fell (16) by the reed (5);
- the breast beam (7) gradually draws off the woven fabric (16) by means of the emery roller (8) and pressure roller (9), and is wound onto the cloth roller (10).

Warp threads are unwound from the warp beam (1) at a speed that maintains a constant average warp tension. All these operations are co-ordinated by the weaving machine.

1.4 Elastic system of fabric formation (ESFF)

The term 'elastic system' denotes the combined span of warp threads and woven fabric, which undergo a certain amount of fluctuation of recoverable stretch deformation during the operation of the weaving machine. During weaving, the warp threads and fabric undergo various deformations: tensile, bending and compressive. Sections of the warp threads from the warp beam ℓ_o (Fig. 1.1) to the cloth-fell ℓ_6 and the fabric $\ell_6, ..., \ell_9$, experiencing deformation under the influence of the weaving machine, form the elastic system of fabric formation (ESFF).

The influence of the main operations of the weaving machine on the different sections of the ESFF varies. For example, when releasing the warp from the warp beam, a reduction of the ESFF deformation (strain) takes place, and when opening the shed, the deformation increases. During the beating-up of the weft into the cloth-fell, the warp threads are under maximum deformation, while the fabric at this point is under minimum deformation.

When there is no contact between the reed and the cloth-fell, the warp threads and the fabric make up a compound flexible body but, during beating-up, the reed disrupts its integrity considerably. The spring-loading of the back-rail serves to compensate for warp tension and, at the same time, due to its deflection, serves as a sensor for the warp let-off mechanism. The take-up roller causes the tension of ESFF to increase due to the withdrawal of the fabric, but the warp let-off counteracts this, and enables a constant mean level of warp tension to be maintained. The configuration and phases of movement of the working mechanisms influence the amount and direction of the deformation of the ESFF elements.

6 Mechanisms of flat weaving technology

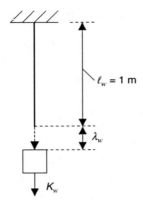

1.2 Method of definition of the stiffness coefficient C_w^1 of a single warp thread.

According to the research of V.A. Gordeev (Ref. 1) (the originator of the theory of ESFF) and his colleagues, it has been shown that, due to the short duration of weaving mechanism operation cycles, deformation of the ESFF elements takes place because of their elastic characteristics. It is possible to make the assumption that tension K_w of the warp threads and tension K_f of the fabric are proportional to the deformation of the warp threads λ_w and of the fabric λ_f (mm), respectively.

To calculate the tension of the warp threads and the fabric, Gordeev (Ref. 1) proposed the use of the *stiffness coefficient* C_w for the warp (and C_f for the fabric), with units of N/mm, which is equal to the ratio of the stretching force K_w to the extension caused by this force. For a single warp thread based on, $\ell_w = 1$ m (Fig. 1.2), the stiffness coefficient C_w^1 is:

$$C_w^1 = \frac{K_w}{\lambda_w} \quad [1.1]$$

where: K_w = single warp thread tension;
λ_w = extension of a single warp thread.

For a unit width (1 m) of the fabric, the stiffness coefficient calculated on the basis of one warp thread can be written as:

$$C_f^1 = \frac{K_f}{\lambda_f \cdot n_w} \quad [1.2]$$

where: K_f = fabric tension;
λ_f = fabric extension;
n_w = the number of warp threads in unit width (1 m).

Introduction: classification and mechanisms of weaving machines

For example, a 1 m length of a single warp thread of average thickness can have the following typical values C_w^1 (Ref. 1 and Ref. 2): for cotton = 0.2 N/mm; wool = 0.1 N/mm; flax = 0.6 N/mm. On the basis of one warp thread, typically, a 1 m length of cotton plain weave fabric will give $C_f^1 = 0.1$ N/mm.

However, the value of the stiffness coefficient will vary according to the length of thread and the fabric. Therefore, in calculating the stiffness coefficients C_{cw} and C_{cf}, it is necessary to take account of the actual length $\ell_{w,f}$ of the thread or the fabric:

$$C_{cw} = \frac{C_w^1}{\ell_{cw}} \quad \text{or} \quad C_{cf} = \frac{C_f^1}{\ell_{cf}} \qquad [1.3]$$

where: C_w^1 = the stiffness coefficient of a 1 m length of single threads;

C_f^1 = the stiffness coefficient of a 1 m length of fabric along warp threads;
ℓ_{cw}, ℓ_{cf} = the length of the tested section of the warp thread or fabric in metres.

For example, for the section of the warp thread in a weaving machine (Fig. 1.1) from the warp beam to the back-rest, $\ell_{cw} = 0.4$ m and the section of fabric from the breast beam to the emery roller $\ell_{cf} = 0.15$ m, we obtain: $C_{cw} = 0.2/0.4 = 0.5$ N/mm, and $C_{cf} = 0.10/0.15 = 0.67$ N/mm. The simplicity and ease of use of definition 'C_c' under given conditions enables the designer to calculate the length of fabric on the weaving machine on the basis of actual technological tensions, according to the type of woven fabric.

Assume that, to calculate the parameters of ESFF, it is necessary to know the resultant stiffness coefficient C_{ESFF} of the different deformation values of the warp threads λ_{cw} and the fabric λ_{cf} that experience the same tension K_c (in the level phase of the shed, for example). The resultant value of stiffness coefficient C_{ESFF} of the *two parts* of ESFF which are connected in series can be given as:

$$\lambda_{ESFF} = \lambda_{cw} + \lambda_{cf};$$

where: λ_{cw} = extension of the warp threads;
λ_{cf} = extension of the fabric.

$$\lambda_{ESFF} = \frac{K_c}{C_{ESFF}} \quad \text{hence,} \quad \frac{K_c}{C_{ESFF}} = \frac{K_c}{C_{cw}} + \frac{K_c}{C_{cf}}, \quad \text{or} \quad \frac{1}{C_{ESFF}} = \frac{1}{C_{cw}} + \frac{1}{C_{cf}},$$

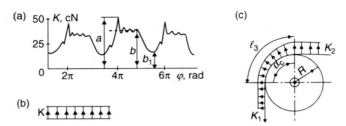

1.3 Warp thread tension on the weaving machine: (a) warp tension at beat-up; (b) warp tension during open shed; (c) thread tension diagram due to frictional contact over stationary back-rest.

where: K_c = tension of the warp threads;
C_{ESFF} = resultant value of stiffness coefficient of the *two parts* of the ESFF;
C_{cw} = the stiffness coefficient of the warp;
C_{cf} = the stiffness coefficient of the fabric;

that is, $$C_{ESFF} = \frac{C_{cw} \times C_{cf}}{C_{cw} + C_{cf}} \qquad [1.4]$$

For example, for C_{cw} = 200 N/mm and C_{cf} = 280 N/mm, we obtain: C_{ESFF} = [(200 × 280)/(200 + 280)] = 116.7 N/mm.

Figure 1.3 shows the typical form of the oscillogram of tension in the warp threads when weaving a fabric of average density, where a = the beat-up tension in the warp, b = warp tension in the open shed during the pick insertion interval; b_1 = warp tension in the closed-shed position.

The tension change ΔK of the elastic system at its total deformation $\lambda_{ESFF} = \lambda_{cw} + \lambda_{cf}$ can be given by the expression:

$$\Delta K = \lambda_{ESFF} \times C_{ESFF} \qquad [1.5]$$

where: λ_{ESFF} = *total deformation of the two parts of the ESFF;*
C_{ESFF} = the resultant value of stiffness coefficient of the two parts of the ESFF.

For example, from λ_{ESFF} = 5 mm and C_{ESFF} = 66.7 N/mm. we obtain: ΔK = 5 × 66.7 = 333.5 N.

If the values of warp stiffness C_{cw} and fabric stiffness C_{cf} are different, the value of ΔK is given by the expression:

Introduction: classification and mechanisms of weaving machines

$$\Delta K = \lambda_{ESFF} \cdot \frac{C_{cw} \times C_{cf}}{C_{cw} + C_{cf}} = \lambda_{cw} \cdot C_{cw} = \lambda_{cf} \cdot C_{cf} \qquad [1.6]$$

where: λ_{ESFF} = total deformation of the two parts of the ESFF;
C_{cw} = the stiffness coefficient for the warp;
C_{cf} = the stiffness coefficient for the fabric;
λ_{cw} = extension of the warp threads;
λ_{cf} = extension of the fabric.

For example, ΔK = 3 mm × 200 N/mm = 2 mm × 300 N/mm = 600 N.

For the variation of *tension* from K_1 to K_2 of a running thread (Fig. 1.3(c)) – for example, around the areas of frictional contact over the fixed back-rest and with *different lengths* of the parts of these branches – the stiffness coefficients C_1 and C_2 are also different. In such cases, the total stiffness coefficient C_{12} for both branches of the warp at the known correlation of tension $n = K_2/K_1$ is given by the formula:

$$C_{12} = \frac{n \cdot C_1 \cdot C_2}{n \cdot C_1 + C_2} \qquad [1.7]$$

where: C_1, C_2 = the stiffness coefficients for the different parts of the branches;
n = the correlation of tension.

For example, for n = 1.6; C_1 = 244 N/mm and C_2 = 484 N/mm, we obtain: C_{12} = (1.6 × 484 × 244)/(1.6 × 484 + 244) = 186 N/mm.

The ESFF elements are under different conditions of stretch. For example, on the parts $\ell_1, \ell_3, \ell_4, \ell_5, \ell_6, \ell_8$ (Fig. 1.1), which are in a relatively free state, it is possible to accept that there is an even spread of deformation along the length (Fig. 1.3(b)) under the influence of tension K. The parts ℓ_0, ℓ_2, ℓ_7 and ℓ_9 experience friction against the warp beam surface, back-rest, breast beam and the emery roller. That is why the thread tension gradually increases from K_1 to K_2, as shown in Fig. 1.3(c). Therefore, it is impermissible to sum the parts of the ESFF with different distribution patterns of λ and K.

1.4.1 The stretch deformation of threads in the let-off zone of the warp beam

Warp threads at the warp beam surface, at any given time during weaving, are in a strain-deformed condition, caused by the winding tension at the formation of the warp beam at the present diameter, and the subsequent thread relaxation due to the removal of the compressive effect of the layers of warp that had been wound outside the present diameter of the beam. The initial (starting) warp threads tension K_{oo} can be more than, equal to or less than tension K_1 of the *warp threads already unwound* depending on the initial tension of the ESFF, which is necessary for forming a given fabric structure.

Figure 1.4 shows the warp let-off zone on the warp beam. Consider an example of inequality $K_1 > K_{oo}$ for most situations of the weaving process. The tension force K_1 of warp threads being unwound extends over a length ($\ell_{oo} + \ell_{co}$), covering an arc of the warp beam in radius ρ over the angle ($\alpha_{wb} + \beta$). Owing to the nature of the warp thread structure, the stretching loads applied, K_1 and K_o, initially cause a relative displacement of fibres along the thread body, a reduction of the angle of twist of separate fibres with respect to the longitudinal axis of the thread and an increase the contact forces between fibres, leading to fibre compression.

The body of the stretched thread on the warp beam can be considered as divided into longitudinal layers of which, after a small preliminary shift, those at the *bottom* remain almost stationary, and those at the *top* continue to undergo displacement, before the moment of balance of forces due to yarn tension causing displacement and resistance of thread elements is reached. The yarn in the boundary between these layers is subject to a complex combination of sliding friction, bending, compression and torsion. The effect of these forces interferes with normal thread elongation. As a result, an area (ℓ_{oo}, α_{wb}) of preliminary displacement of thread elements is formed. Any further increase in warp tension following after this area may cause the thread body to slide relative to the warp beam surface (ℓ_{co}, β).

The increase in tension from K_{oo} to K_o of a yarn element in the leading area of displacement can be expressed by using Euler's law:

$$K_o = K_{oo} \times \exp(\mu \cdot \alpha_{wb}), \qquad [1.8]$$

where

$\mu = f + (v \cdot \rho \cdot \alpha_{wb})$ the coefficient of tangential resistance to longitudinal thread sliding;

α_{wb} = an angle of the area of preliminary displacement of threads elements;

ρ = the warp beam radius;

v = the coefficient of coupling of a thread with the surface of the warp beam, ratio carried to unit of threads length;

f = the coefficient of sliding friction which is independent of thread length.

Introduction: classification and mechanisms of weaving machines

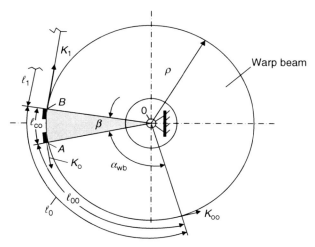

1.4 Warp let-off zone on the warp beam.

The analysis of the thread sliding zone (ℓ_{co}, β) gives the following expression for K_0:

$$K_0 = \frac{K_1}{(f \cdot \sin\beta + \cos\beta)} \quad [1.9]$$

where: K_1 = the tension of a warp thread;
β = the angle of sliding of the thread body relatively to the warp beam surface;
f = coefficient of sliding friction which is independent of thread length.

By equating the first derivative $dK_0/d\beta$ to zero:

$$\frac{dK_0}{d\beta} = -\frac{K_1(f \cdot \cos\beta - \sin\beta)}{(f \cdot \sin\beta + \cos\beta)^2} = 0 \quad [1.10]$$

where: K_1 = the tension of the warp thread.

We find the critical value of the angle β, defining the extent of the sliding zone (ℓ_1, β):

$$\beta = \text{arctg } f \quad [1.11]$$

12 Mechanisms of flat weaving technology

From the equality of expressions [1.8] and [1.9], it is possible to define the main parameter of the preliminary displacement area – limiting value of angle α_{wb}:

$$\alpha_{wb} = \frac{-f + \sqrt{f^2 + 4v\rho \cdot \ell n S}}{2v\rho} \qquad [1.12]$$

where: f = the coefficient of sliding friction which is independent of thread length;
α_{wb} = the angle of the zone of the preliminary displacement of thread elements;
v = the coefficient of frictional coupling of a thread with the surface of the warp beam, ratio carried to unit of thread length;
ρ = the warp beam radius;
S = coefficient: $S = \dfrac{K_1}{K_{\infty} \times (f \cdot \sin\beta + \cos\beta)}$
β = angle of sliding.

For example, for $f = 0.306$; $v = 0.35$; $K_{\infty} = 0.12$ N; $K_1 = 0.25$ N; $\rho = 200$ mm; we obtain: $\beta = 17°$, $\alpha_{wb} = 95°$.

Deformation by stretching of the part ℓ_{co} of threads in the unwinding area of the warp beam is given by the integral:

$$\lambda_{\ell_{co}} = \frac{K_1 \cdot \rho}{C_w^1} \int_0^{\beta} \frac{d\gamma}{f \cdot \sin\gamma + \cos\gamma}, \qquad [1.13]$$

where: K_1 = tension force of warp thread;
ρ = the warp beam radius;
C_w^1 = the stiffness coefficient of 1 m of warp thread;
γ = the angle, which varies from 0 to β;
f = the coefficient of sliding friction which is independent of thread length;
β = the angle of sliding of the thread body relative to the warp beam surface.

The solution of [1.13] gives:

$$\lambda_{\ell_{co}} = \frac{K_1 \cdot \rho}{C_w^1}\left(\frac{1}{a}\ell n\left|\frac{a - f + tg(\beta/2)}{a + f - tg(\beta/2)}\right| + C\right), \qquad [1.14]$$

Introduction: classification and mechanisms of weaving machines

where: K_1 = the tension force of warp thread;
ρ = the warp beam radius;
C_w^1 = the stiffness coefficient of 1m of warp thread;
a – coefficient: $a = \sqrt{1+f^2}$;
f = the coefficient of sliding friction which is independent of thread length;
β = the angle of sliding of the thread body relative to the warp beam surface;
C = the constant of integration.

The constant of integration C will be obtained from the boundary conditions as:

$$C = -\frac{\ell_{co}}{\rho} - \frac{1}{a} \ell n \left| \frac{a-f}{a+f} \right| \qquad [1.15]$$

where: ℓ_{co} = the length of unwound warp threads;
ρ = the warp beam radius;
a = the coefficient, where $a = \sqrt{1+f^2}$;
f = the coefficient of sliding friction which is independent of thread length.

For example, for $f = 0.42$; $\rho = 200$ mm; $C_w^1 = 0.2$ N/mm; $K_1 = 0.25$ N; $\beta = 17°$; $\ell_{co} = \pi * \rho * \beta/180° = 59.34$ mm. We obtain: $a = \sqrt{1+0,42^2} = 1.085$;

$$C = -\frac{59.34}{200} - \frac{1}{1.085} \ell n \left| \frac{1.085 - 0.42}{1,085 + 0.42} \right| = 0.4558$$

and

$$\lambda_{\ell_{co}} = \left| \frac{0.25 \cdot 200}{0.2} \left(\frac{1}{1.085} \ell n \left| \frac{1.085 - 0.42 + 0.1495}{1.085 + 0.42 - 0.1495} \right| + 0.4558 \right) \right| = 3.3 \text{ mm}$$

1.4.2 Equivalent thread and fabric length in the ESFF

A solution to give the warp and fabric length in the ESFF was proposed by Gordeev (Ref. 1), who introduced the notion *equivalent thread length* L_{we}. L_{we} is considered to be the *conventional* thread length in a *free* condition (without friction) that has received the same *total* stretching deformation of as the real section of thread, experiencing friction under the influence of K_2, the tension of the leading branch (as shown in Fig. 1.3).

For the warp beam, the EQUIVALENT PART L_{wc} (metre) of section ℓ_0 (Fig. 1.1) of the warp thread for the given radius of warp beam ρ, friction coefficient f_w, initial thread tension K_0 and the leading branch tension K_1 will be given by the expression, m:

$$L_{wc} = \frac{\rho}{f_w}\left(1 - \frac{K_0}{K_1}\right), \qquad [1.16]$$

where: ρ = the warp beam radius;
f_w = the coefficient of sliding friction which is independent of thread length;
K_1 = the tension of the warp thread being unwound;
K_0 = the initial warp thread tension.

For example, for $f = 0.2$; $\rho = 200$ mm; $K_0 = 0.07$ N; $K_1 = 0.25$ N; we obtain:

$$L_{wc} = \frac{200}{0.42}\left(1 - \frac{0.07}{0.25}\right) = 342.8 \text{ mm}$$

For the back-rest with a radius R (Fig. 1.3(c)) and angle of warp thread α_c, ℓ_2 can be replaced by:

$$L_{2e} = \frac{R}{f_w}\left(1 - \frac{1}{e^{f_w \cdot \alpha_c}}\right) \qquad [1.17]$$

where: R = the back-rest radius;
f_w = the coefficient of sliding friction which is independent of thread length;
α_c = the angle of warp thread, of ℓ_2.

For example, for $R = 66.5$ mm; $f = 0.25$; $\alpha_c = 90° = 1.57$ rad; $\ell_2 = \pi * 66.5 * 90°/180° = 104.4$ mm, we obtain:

$$L_{2e} = \frac{66.5}{0.25}\left(1 - \frac{1}{e^{0.25 \times 1.57}}\right) = 86.3 \text{ mm}$$

According to these two formulae, it is possible to define the equivalent length of the fabric on ℓ_7 and ℓ_9, correspondingly.

Introduction: classification and mechanisms of weaving machines

As a result, calculating length of the warp threads in the ESFF will be, m:

$$L_w = L_{we} + \ell_1 + L_{2e} + \ell_3 + \ell_4 + \ell_5, \qquad [1.18]$$

where: L_{we} = the equivalent part (in metres) of ℓ_0;
L_{2e} = the equivalent part of ℓ_2;
$\ell_1, \ell_2, \ell_3, \ell_4, \ell_5$ = sections of warp threads in the ESFF.

For example (see Fig. 1.1), for the Sulzer weaving machine: $\ell_1 = 400$ mm; $\ell_3 = 475$ mm; $\ell_4 = 309$ mm; $\ell_5 = 137$ mm; $L_{we} = 342.8$ mm; $L_{2e} = 86.3$ mm we obtain: $L_w = 342.8 + 400 + 86.3 + 475 + 309 + 137 = 1750.1$ mm $= 1.75$ m. Calculating length of the fabric in the ESFF is carried out by analogy with warp threads according to the construction of the weaving machine.

1.5 Advantages and disadvantages of different weaving machines

- Shuttle weaving machines can manufacture most varieties of woven fabric; however, their operation is relatively slow (up to 250 rpm).
- Shuttleless weaving machines use rapier, projectile, air, water, electromagnetic or weft inertia based methods of weft insertion. Pneumatic and hydraulic (water-jet) methods produce higher weft insertion rates. Rapier and other mechanisms are relatively slower.

1.6 Questions for self-assessment

1. What are the different ways in which weaving machines can be classified?
2. On the basis of what characteristics are shuttle and shuttleless weaving machines classified?
3. What are the methods of shedding used on weaving machines?
4. What are the different methods of picking used on weaving machines?
5. How are weaving machines normally identified by manufacturers?
6. What are the basic mechanisms of weaving machines?
7. What is purpose of each basic mechanism of a weaving machine?
8. What is the 'elastic system of fabric formation (ESFF)'?
9. What are boundaries (limits) of the ESFF on a weaving machine?
10. What is the extent of the ESFF on a weaving machine?
11. What are the specific features of the separate parts of the ESFF?

12. How does the diagram of distribution of thread or fabric tension at a free component essentially differ from that on the surface of a guiding component (Fig. 1.3)?
13. What is meant by the term 'stiffness coefficient'? How is the stiffness coefficient of a thread defined?
14. Why it is necessary to define the stiffness coefficient of a fabric based on one warp thread, but counting the entirety of the fabric width?
15. What is the relationship between the stiffness coefficient of a 1 m piece of thread (or fabric) and a piece of any given length?
16. What is meant by 'the equivalent length of a thread'? What are the factors that define its magnitude?
17. What are the elements which define the length of the ESFF of a weaving machine?

1.7 References

1. Gordeev V.A. and Volkov P.V., 'Weaving', Leg. and Pitsh. Prom., Moscow, 1984 (in Russian).
2. Choogin V.V., Kahramanova L.F. and Nedovisiy M.N., 'Technology of Weaving Manufacture', State Technical University, Kherson, 2008 (in Russian).

2
Mechanisms of the weaving machine for warp release and warp tension control

DOI: 10.1533/9780857097859.17

Abstract: This chapter deals with the concept of warp release (*warp let-off*), and the mechanisms concerned with the weaving process. Both manual and automatic warp beam brakes are presented. First, a description is given for how the required length of warp is *released* in into the weaving area, rather than simply being *unwound* from the warp beam. This also proves the feasibility of the operation of the beam regulator which handles equal and unequal thread tension coming from two warp beams on one weaving machine by means of the differential. The chapter closes with an example of the comparative analysis of all release mechanisms and, also, recommendations on how to use them.

Key words: warp release, beam regulator, beam brakes.

2.1 Introduction: mechanisms of the weaving machine for warp release and warp tension control

In the process of forming fabric, it is necessary to take up a freshly woven section of fabric in each weaving cycle and release a corresponding length of warp, under the necessary tension, into the weaving area from the warp beam (Ref. 1). The length of warp released, $\Delta \ell_{wp}$, for each rotation of the main shaft of the weaving machine should be more than the length of freshly woven fabric, $\Delta \ell_f$, taken up from the fabric forming area because of the crimping of the warp threads in the fabric a_{wp}:

$$\Delta \ell_{wp} = \frac{1}{P_{wft}}\left(1 + \frac{a_{wp}}{100}\right), \qquad [2.1]$$

where: P_{wft} = weft density, threads/cm;
a_{wp} = warp crimp in the fabric (%).

For example, if P_{wft} = 20 threads/cm; a_{wp} = 8%; we obtain:

$$\Delta \ell_{wp} = \frac{1}{20}\left(1 + \frac{8}{100}\right) = 0.054 \text{ cm} = 0.54 \text{ mm}.$$

In practice, there are two concepts which can cause some confusion: warp 'release' and warp 'delivery' (or feeding) to the ESFF (Ref. 2). On a weaving machine, the warp is always under tension K which creates turning moment $M_m = K * \rho$ on the warp beam according to the radius ρ of yarn wound on it. Under the influence of this turning moment, the warp beam is always taut since the warp release mechanism restrains its rotation. This eliminates all backlash (free movement) between the warp beam drive gear wheels and the restraining mechanism (e.g. a worm gear).

The restraining mechanism is controlled by a warp tension sensor, which operates within set parameters, in such a way as to enable rotation of the warp beam to slacken precisely the right length of the warp to maintain warp tension at a constant average value as required in a given particular weaving operation. In contrast, warp *delivery* (or feeding) would involve the release of warp of a fixed length each time, by carrying out warp beam rotation according to set parameters.

Warp feeding is used only on special (e.g. shaggy) types of weaving machines which require periodic mandatory warp let-off during the formation of pile or similar features under a reduced level of warp tension. On wide, high-speed weaving machines with a large warp beam of considerable inertia, the combination of both processes is possible: there is initial warp feeding followed by a controlled release.

In the normal process of weaving a fabric of a stable structure, it is necessary to maintain a cyclic MODE of tension variation in the ESFF. For this purpose, first, warp threads on a weaving machine should have a primary optimum level tension setting during fabric formation in the weaving cycle. The likelihood of thread breakage increases if the warp tension is too high or too low. If the primary warp tension setting is too high, the dynamic tension of threads increases during shedding and especially on the beating-up of weft threads, which may lead to their breaking strength being exceeded. If the tension setting is too low, there is the possibility of warp threads becoming tangled in the shed and the droppers sagging, which would also increase the probability thread breakage.

The optimum primary setting of the warp tension level depends on many factors, in particular, the warp and the weft sett, the degree of interlacing between warp and weft, the type of fibre in the warp and weft and the type of weaving machine. For light fabrics, the normal primary warp tension setting is from 5 to 15 cN/thread; for average fabrics, from 15 to 50 cN/thread; for closely woven (highly set) fabrics, from 50 to 500 cN/thread; for heavy-weight fabrics, from 500 to 1500 cN/thread. Maintaining an optimum primary warp tension setting in the ESFF is achieved by means of warp beam brakes or regulators.

Figure 2.1 shows the basic design for obtaining the optimum primary warp tension setting in the ESFF. Two counteracting tuning moments are

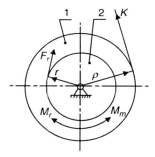

2.1 The forces operating on warp beam. Refer to text for detailed explanation of components.

shown: the driving moment M_m from the action of the warp threads tension K; and the moment of resistance M_r arising from the action of the resisting force F_r. Force K, due to the tension of the running branch of the warp, is applied at the radius ρ of the warp threads on the beam, which gradually decreases during the unwinding of the warp beam 1. The resisting force F_r is applied against the rotation of the warp beam at radius r of the brake pulley 2 or the warp beam gear wheel. The general tension of the running warp line K results from its mean (average) K_s and dynamic K_d components. The condition of balance for the warp beam satisfies the following relationship:

$$K_s \cdot \rho = F_r \cdot r, \qquad [2.2]$$

where: K_s = the average component of tension of the running warp line;
ρ = the radius of the warp on the beam;
F_r = the resisting force;
r = the radius of the brake pulley.

It is obvious that with continued weaving, due to the release of the warp, the beam radius, ρ, gradually diminishes and the mean tension K_s will increase as given by:

$$K_s = F_r \cdot \frac{r}{\rho}, \qquad [2.3]$$

where: K_s = the mean component of tension of the running warp line;
F_r = the resisting force;
r = the radius of the brake pulley;
ρ = the radius of the warp threads on the beam.

Hence, the general tension K will gradually increase. Therefore, to maintain the required primary warp tension setting, the warp release mechanism should maintain a mean level of tension K and amplitudes ΔK of its cyclic fluctuation as radius ρ diminishes with continued weaving.

2.2 Warp brakes

The extent to which the warp is released by the brakes is governed by the excess of the moving turning moment M_m over the moment of resistance M_r (Ref. 2). In practice, two types of warp brakes are used to restrict the rotation of the warp beam: those requiring manual control; or those automatically controlled by the resistive moment M_r, for opposing the rotation of the warp beam (Ref. 1).

The force of resistance F_r to the rotation of the warp beam by the warp brake is created by the effective force of friction T applied to a brake pulley (1) (Fig. 2.2). There are various ways to achieve this: mechanical friction against a band (2) (Fig. 2.2(a)), or, in addition, against a brake shoe (5) (Fig. 2.2(b)) with the sum of forces T_1 and T_2.

Figure 2.2(a) presents a diagram of a band brake in which the resistance to rotation of the warp beam (3) is created by the force of friction T of a pulley (1) against the band (2). The force of friction T is created by the effort of the band (2) on contact with the brake pulley (1). This effort depends on the tension of a tape (2) – created, for example, by weight G and weighted lever 4. The tension of tape Q can also be created in other ways, for example, with a spring. In this brake, for the maintenance of equality $M_m = M_r$ while ρ gradually diminishes it is possible to increase force Q by reduction of the length of the shoulder ℓ_2.

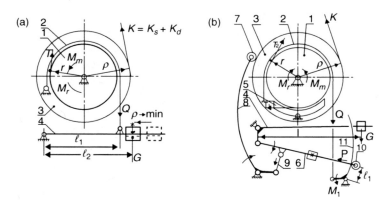

2.2 Friction brakes. (a) Tape with manual regulation of braking force; (b) combined differential automatic (Hartmann).

Thus, for the regulation of tension K in the process of unwinding the warp beam, it is necessary to *manually* shift the weight (G) along the lever (4). A step change in K_s corresponds to discrete regulation of M_r. The quality of the adjustment – that is, the uniformity and timeliness of regulation K_s – is dependant on the weaver's experience.

In the case of automatic warp brakes, the automatic differential combined brake of the shuttle weaving machine (Hartmann), as shown by Fig. 2.2(b), is well-known. Here, braking of the warp beam (3) is achieved by the action of a brake band (2) and brake shoe (5) on the brake pulley (1). The force of friction T_1 is mostly created by the weight of the warp beam and, to a lesser degree, by the pressure of the band (2) on the pulley (1). In the process of unwinding the warp, the reduction of pressure of the pulley (1) on brake shoe (5) causes force T_1 to decrease as well.

The force of friction T_2, as in the case of the usual band brake (Fig. 2.2(a)), depends on the tension of band Q. However, in this mechanism, tension Q decreases automatically in the process of the reduction of radius ρ of the warp beam (3). As the diameter of the warp beam decreases, the roller (7) of the diameter gauge (8) lowers the control rod (6) by means of a link (9). Thus, the roller (10) applies pressure P upon the differential lever (11) with a smaller shoulder ℓ_1. Thus, a decrease in the turning moment M_1 leads to a reduction of the magnitude of Q. The initial level of K_s is manually set by the length of the shoulder ℓ on the lever (4), with the weight G.

The profile of the differential lever (11) is determined, based on the calculation for the preservation of the fluctuation level only of the *static* component of the tension of warp K_s. Therefore, in practice, the tension K increases as the warp beam is unwound, leading to an increased breakage rate of warp threads. Dynamic component K_d (Ref. 1) of the general tension K depends on the moment of inertia I of the warp beam and the angle of rotation φ of the warp beam during the release phase in time t by radius ρ of the warp beam:

$$K_d = f(I,\rho,\varphi,t) \qquad [2.4]$$

where: I = the moment of inertia of the warp beam;
ρ = the radius of the warp threads on the beam;
φ = the angle of rotation of the warp beam;
t = time.

Considering component K_d in the given design of the brake, additional compensator links are included to reduce the tension Q of the band (2) in the process of unwinding the warp beam. Basically, this problem can be resolved by changing the setting contour of the differential lever (11). However, the

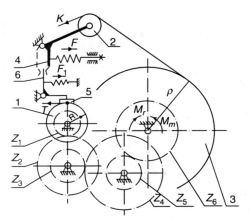

2.3 An automatic brake for a warp beam with a reducer on a pneumatic weaving machine (Elitex).

compensators cannot solve the problem of stabilization of tension K for multiple fabric structures.

Structurally, the pulley (1) (Fig. 2.2) diameter is approximately two or three times the diameter d of the tube of the warp beam (3). Therefore, it is necessary to satisfy the condition $T \gg K$ at full warp beam (ρ_{max}) for its control before the beginning of release of the warp. Having equal diameters of the warp beam and the brake pulley ($\rho = r$) is only sufficient to provide equal forces T and K. As the warp beam is used up, when $\rho \ll r$, it is necessary to provide for an inequality of forces of opposing sense according to $T \ll K$.

The air-jet (pneumatic) weaving machine (Elitex) has a highly refined automatic brake with a reducer and a feedback arrangement (Fig. 2.3). In this mechanism, the brake pulley (1) is mounted separately from the warp beam (3). The pulley (1) and warp beam (3) are connected by means of geared three-step reducers, which have teeth numbering $Z_1, \ldots, Z_6$. This method has permitted considerable lowering of the necessary magnitude of friction force T to maintain the set moment of resistance M_r. At the band brake (Fig. 2.2(a)), the friction force is:

$$T = K_s \cdot \frac{\rho}{r}, \qquad [2.5]$$

where: K_s = the static components of tension of the running warp line;
r = the radius of brake pulley;
ρ = the radius of the warp threads on the beam.

For example, for $K_s = 400$ N; $\rho = 200$ mm; $r = 150$ mm; we obtain:

$$T = 400 \cdot \frac{200}{150} = 533,3 \text{ N}$$

Then, in Fig. 2.3, it follows that: $T = K_s \cdot \dfrac{\rho}{R \cdot i}$, [2.6]

where: K_s = the static components of tension of the running warp line;
ρ = the radius of the warp threads on the beam;
R = the radius of the brake pulley;
i = the transmission ratio.

For example, for $i = 100$ (for the Elitex weaving machine) and $r = R$, we obtain:

$$T = 400 \cdot \frac{150}{150 \times 100} = 4 \text{ N}.$$

The transmission ratio is:

$$i = \frac{Z_2 \cdot Z_4 \cdot Z_6}{Z_1 \cdot Z_3 \cdot Z_5}, \quad [2.7]$$

where: $Z_1 - Z_6$ = numbers of teeth on the gear wheels.

The numbers of teeth in the gears may vary. Thus, the value of 'i' can range from $i = 100$ to $i = 170$. For example, in one variant: $Z_1 = 18$; $Z_2 = 100$; $Z_3 = 21$; $Z_4 = 100$; $Z_5 = 21$; $Z_6 = 130$; $i = 163.77$.

In order to avoid an undesirable change of the senses of inequality of forces T and K in the process of the reduction of radius ρ of the warp beam, it is desirable to develop brakes with reducers that allow a lower force of braking T with a small diameter of a brake pulley (1) in comparison with the diameter of a warp beam tube. In producing light fabrics, K is obtained by a spring F_1 of low resistance. In this mechanism, the more resistant spring F achieves equilibrium against the force exerted by the two spans of warp threads acting on the back rest (2). If the tension K increases, the dual shoulder lever (4) of the responsive mechanism of the back-rest (2) will press the bottom shoulder to the left on the vertical shoulder of the dual shoulder brake lever (6); thus, brake shoe (5) will apply a reduced pressure on the brake friction pulley (1). The reduction of the force of friction T will allow the warp beam (3), under the influence of warp tension K, to turn over a larger angle and to release a longer length of warp. The ESFF tension will decrease to its normal level.

In producing fabrics of higher density (high tension of all threads in the fabric structure) the dual shoulder lever (4) constantly presses the lever (6) (shown dashed) from the left to the right. In this case, the more resistant spring F, together with spring F_1, create a normal pressure level of the brake shoe (5) on the brake pulley (1). Spring F_1 provides only a minimum level of friction force of T. Decreasing the ESFF tension from its normal level causes the back-rest (2) to oscillate at a higher position. In this case, spring F will have a greater influence on the shoulder (6), and the brake shoe (5) will press more firmly on the pulley. An increase of friction force T will lead to a reduction in the release of the warp threads from the warp beam, and this will raise the level of fluctuation of tension K to its normal value.

Alternatively, the moment of resistance to the rotation of the warp beam can be achieved by using an electromagnetic system, which is considered a superior method. In this type of system, an electromagnetic field controls the operation of the release pulley by means of either a sensor operated from the back-rest or a special warp tension sensor.

2.3 Warp regulators

As with brakes, the primary purpose of warp regulators is to to control the release of the warp beam, thereby governing the release of warp and managing warp tension in the ESFF, and. The basic difference between a regulator and a brake is that a brake provides *passive* resistance to the moving moment M_m with respect to the axis of the warp beam. A regulator provides *active* resistance to M_m by means of a specific mechanism with a drive that is independent from the main shaft of the weaving machine (Ref. 2).

Warp regulators should satisfy the following requirements:

- The amount of warp released from the warp beam should correspond to the consumption of warp in the formation of *each element* of the fabric in each cycle of thread interlacings.
- The total amount of warp threads released for one repeat of the fabric interlacing cycle should remain *constant* for the whole period during which the warp beam is unwound.
- To allow for a wider assortment of fabrics, it is necessary to have a significant range of stable regulation options from which to choose the amount warp release and warp tension values.
- The sensitivity of a regulator to the change of release and warp tension should be sufficient to form a stable fabric structure and, simultaneously, regulate over a wide range of pickspacing settings.

The constructional peculiarity of all warp release regulators is the need to turn the warp beam (on each revolution of the main shaft of the weaving machine) through a very small angle (range: 5–60 min. of arc). Therefore, in regulators with high transmission ratios (1:200, ... ,1:600), either worm or planetary gear reduction mechanisms are used. A further benefit of these types of reduction mechanisms is that they obviate any undesirable excess of moment M_m in the warp beam movement over the maximum moment M_r which the regulator can safely withstand.

To understand the basic performance of any warp regulating mechanism correctly, it is necessary to establish, first, the precise location where the forces meet and to achieve equilibrium of the *two resultant forces*: from the tension of a run (or branch) of warp threads, K_{ct}, and from the drive (gear) of the mechanism of a regulator, T_{ct}. Then, it is necessary to define the operational principles of their interaction, to understand the process of cyclic and random deviations from the mechanism's state of equilibrium. Figure 2.4 presents an illustration of a worm gear warp regulator the function of which is independent of the thread tension mechanism used in shuttle weaving machines. A worm gear warp regulator functions automatically, based only on the decreasing diameter of the warp beam (Ref. 2).

Warp release comes about through the *joint action* of the mechanism and the ESFF tension. The crank of the main shaft (1) drives a connecting rod

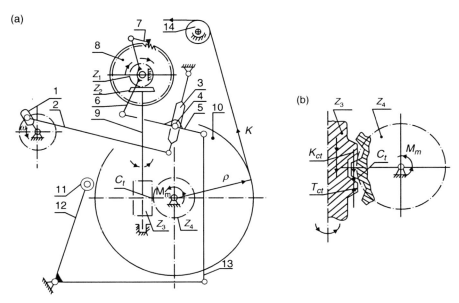

2.4 Worm warp regulator independent of thread tension in action (a) the mechanism; (b) interaction of teeth of gear wheels.

(2), from which a link (3) receives an oscillatory movement. From this lever, with the help of the link hinge (4), traction arm (5), dual shoulder lever (6) and ratchet (7), a ratchet wheel (8) is exposed to periodic motion. The ratchet wheel is turned by means of conical gear wheels Z_1, Z_2, A worm shaft (9) drives the worm Z_3. The tension of threads K creates turning moment M_m, which acts on the worm gear wheel Z_4.

The gear wheel tooth at point C_t (Fig. 2.4(b)) presses against a tooth of the worm Z_3 with a force K_{ct} which resists that action with a force T_{ct}. These two forces are equal at any time. It should be noted that the worm can resist turning against a force even greater than the maximum value that K_{ct} is capable of reaching. So, gear Z_4 can only turn when the worm rotates, but not vice versa. Therefore, the rotation of worm Z_3 allows gear wheel Z_4 to have the freedom to move under the influence of turning moment M_m. The resulting unwinding of the warp beam (10) slackens the warp threads.

Automatic control of the amount of warp released is carried out by means of the core mechanism of the warp beam gauge, which consists of the roller (11), the lever (12) and the vertical rod (13) which lowers sliding block (4) along the link lever (3). This increases the stroke of the catches (7) and, hence, the angle of rotation of the warp beam (10). The back-rest (14) is not involved in the action of the warp release regulator. In this type of regulator, there is *no feedback* based on warp tension and the release mechanism; therefore, the amount of warp released is governed by the initial adjustment of the length of crank (1). The regulator shown in Fig. 2.4(b) is used on wide shuttle weaving machines.

Figure 2.5 presents one of the widely used iterative regulators in shuttle weaving: the Röper regulator. This regulator has a planetary reducer whose operation is *dependent* on the warp diameter *and* the tension of the warp yarns (Ref. 1). The drive for the regulator is obtained from a ratchet mechanism which is driven by the sley, and regulator action is controlled by a feedback mechanism based on a responsive back-rest. The Diederichs and the Northrop weaving machines use this arrangement. The turning moment M_m influences the movement of the warp beam (1) (Fig. 2.5(a) and (b)), by means of gear wheels Z_3 and Z_2, to turn the shaft (2) through static ratchet (3). However, its movement locks against a lateral tooth (4) on the sun gear wheel of the planetary reducer (5). As a result, at point C_t (Fig. 2.5(c) and (d)) there are two forces: K_{ct} and T_{ct}. However, the warp beam (1) cannot turn under the influence of the tension of warp threads K.

Only following the movement of the sley 7 (Fig. 2.5) can the connecting rod (8) be moved away from the warp beam by a pin (6). A lever (9), together with the bottom-side link lever (10), turn the ratchet wheel Z_1 by means of another ratchet (11) through a set degree. Ratchet wheel Z_1 will turn the gear wheels of the planetary regulator (5), which carry the *lateral*

Mechanisms for warp release and warp tension control 27

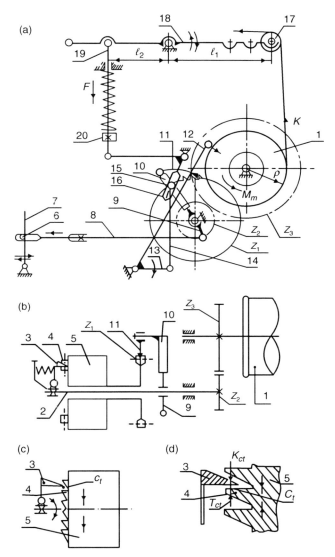

2.5 Planetary warp regulator of action dependent on tension (Röper).
(a) The mechanism; (b) communication of a ratchet with warp beam;
(c), (d) contact of lateral teeth with the ratchet on intermediate shaft.

teeth (4). Only in this case – after the tooth (4) moves in the same direction as the catch (3), and together with it everything which is in a state of tension – can the system can move: shaft (2), gear wheels Z_2 and Z_3 and, hence, the warp beam (1). This will release warp from the warp beam. Thus, *under no circumstances* can the planetary reducer rotate the warp beam by itself (due to the direction of its profile, the tooth (4) cannot apply a driving force to the other gear tooth (3)).

By means of its bottom arm (13) and link (14), the warp beam diameter gauge (12) moves the sliding block (15) downwards as the beam radius ρ diminishes. The top link lever (16), which is under the control of forces F and K, moves the top link lever (10) to the left if there is an increase in warp tension. This moves a ratchet (11) backwards on the teeth of ratchet wheel Z_1, which has the effect of increasing the stroke of the ratchet (11). Ratchet wheel Z_1 then turns the planetary gear wheels of the reducer (5) through an increased arc, increasing the stroke of the lateral teeth (4) and, together with them, the movement of the catch (3), rotation of the shaft (2), gear wheels Z_2, Z_3, and the rotation of the warp beam (1). Thus, the length of the released warp remains the same as ρ diminishes.

Corrections to the amount of warp released in response to a random deviation of the amplitude of fluctuation of the ESFF tension from its set value are made through the responsive mechanism of the back-rest. Here, the back-rest (17) carries out two functions: detecting what forces are at play and gauging the change of tension K in the warp threads. The back-rest (17) can be placed in one of three notches in a lever (18), connected by a pivot to the vertical link (19) and loaded with spring F, applying pressure on a nut (20).

The link (19) is connected to the horizontal arm of the top link lever (16). The top link lever, by means of the sliding block (15), is connected with the bottom link lever (10). With increasing warp tension, the back-rest (17) falls and the left arm of the lever (18) rises and, when the traction rod (19) turns the two-shoulder lever of the top link lever (16) clockwise, the bottom link lever (10) is displaced to the left. Pawl 11 slides on the teeth of ratchet Z_1, increasing rotation angle of gear Z_1. As a result of this, the bottom shoulder (9) of the bottom link lever (10) will be displaced some considerable way to the right, thereby reducing the idle course of the finger (6) of the sley blade (7), and moving in the eye of the link lever (draught) (8).

As a result, during a movement of the sley in the front position, the finger (6) of the sley blades meets an internal surface of an eye (8) earlier and will move a greater distance to the left; this leads to an increase in the angle of rotation of the bottom link lever (10) and ratchet Z_1 due to the movement of the Pawl (11). The amount of warp released will be increased, and the tension of the warp threads decreases to their former level.

The responsive mechanism of the back-rest (17) makes cyclic fluctuations according to a set mode. A condition of balance of this system is the equality of the moments due to the action of two forces: spring force F and the opposing tension of the warp passing over the back-rest. Regulating shoulders of length ℓ_1 and ℓ_2, and also the force of compression of the spring F, make it possible to establish a set level of tension K_y. For producing a

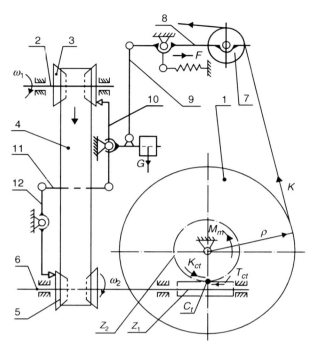

2.6 Warp regulator with belt type variator with continuous warp release.

more closely woven fabric structure, it is necessary to reduce ℓ_1 and to increase ℓ_2 or F.

If, instead of a planetary reducer being used as the controlling mechanism, a worm is used, because of its more rigid construction, while all else remains the same, it is possible to produce a fabric of higher weft density by a given weft yarn and to reduce the size of the cloth fell movement when beating up (i.e. the size of deformation or elongation of the warp threads at the beating-up of weft threads). Such regulators are used on narrow shuttle weaving machines for the development of fabrics of small and average intensity. Figure 2.6 presents a diagram of a warp regulator (Hunt type) with a belt (frictional) variator and with continuous release of the threads, dependent on the tension of the warp threads (Ref. 2).

Here, a worm Z_1 controls the movement of the warp beam (1) through the warp beam gear wheel Z_2 under the influence of moving moment M_m. The rotation of the main shaft of the weaving machine drives the shaft (2) and, hence, the top cones (3) of the frictional variator. The rotation of the top cones is transferred to the shaft (6) and to the worm Z_1 by means of the belt (4) and the bottom cones (5). Thus, continuous rotation of the worm Z_1

from the main shaft of the weaving machine allows gear wheel Z_2 and the warp beam (1) to have continuous movement under the influence of the tension K.

The teeth of gear wheel Z_2 engage with the teeth of the worm Z_1 at point C_t with the force K_{ct} resulting from warp tension K. In case of deviation of the back-rest (7) from the set extremes of fluctuation due to changed tension of warp threads, the mechanism responds so as the restore the level of warp tension. For example, if the warp tension increases, through the system of levers (8), (9), (10), (11), (12), the top cones (3) converge and the bottom cones (5) diverge. As a result, the speed of rotation of the shaft (6) and worm Z_1 increases; hence, the speed of the worm wheel Z_2 increases. Thus, the warp beam (1) releases an increased length of warp into the working area, and the ESFF tension will decrease to normal. The principal weakness of the Hunt regulator is the need to gradually lower the back-rest (7) during the process of reducing ρ, to increase the angle of rotation of the warp beam (1) and to maintain the length of warp thread released. However, this can only be achieved with a gradual increase in tension K. The Hunt regulator is used in shuttleless weaving machines.

Figure 2.7 presents a hydraulic warp regulator, the function of which is also based on the tension of warp threads. The driving moment M_m creates the resulting force K_{ct} on the teeth of worm gear wheel Z_6 at point C_t engaged with the teeth of worm Z_5. The rotation of the worm Z_5 is carried out by means of a shaft (2), through gear wheels Z_4, Z_3, Z_2, Z_1, from the rotary hydraulic motor (3) and is operated by a pressure head of a fluid (oil) supplied from a rotary gear pump. The regulating hydraulic valve (4) governs the speed at which the rotor of the hydraulic motor (3) operates. The regulating hydraulic valve is operated by the back-rest (5) by means of the dual shoulder lever (6). A change in the tension of warp threads K (for example, by an increase from the normal setting) will cause the back-rest (5) to lower and the rod (7) will be moved to the left by the shoulder (6).

Thus, the spool of the valve (4) opens the line (8) of oil flow into the axial piston hydraulic motor (3), whose rotor speed increases, leading the worm Z_5 to rotate more rapidly. As a result, the gear wheel Z_6 and the warp beam (1) can turn at a higher rate and release a longer length of the warp. The tension K will thus decrease to the normal level. As in the Hunt regulator mechanism, there is no warp beam diameter gauge. Therefore, to ensure a constant release of warp during the take-up of the warp threads, it is necessary to increase the rotation angle of the warp beam. To achieve this, the back-rest should fall gradually, which is possible if the tension of warp threads increases. Such regulators are used on shuttleless hydraulic weaving machines.

Two or three individual warp beams are used on wide weaving machines with a reed width over 2.5 m. The peculiarity of such widths is the

Mechanisms for warp release and warp tension control

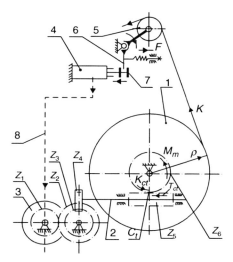

2.7 Hydraulic warp regulator.

impracticability of winding threads on *every* warp beam to carry an identical length of warp. Therefore, once a warp beam is depleted, a considerable length (4, ... , 30 m) of warp on one (or two) warp beams remains unused and has to be discarded. At the high cost of raw materials (such as wool), these remnants increase the cost price of the fabric. Further, it is necessary to consider the phenomenon of ESFF 'individualization' of every system on a weaving machine due to the collective effect of small differences in the machine components and their operation. This can cause differences of deformation and tension of yarn on the warp beam, and the tension of warp and weft threads in the fabric, resulting in different degrees of shrinkage of the warp and weft in the fabric, with negative consequences on fabric quality.

The frictional warp regulator, known as the 'differential' (found on Sulzer (Ref. 2) wide weaving machines), has been adopted the most in wide weaving machines to develop two, three or more lengths of fabric (using one or two warp beams) on one weaving machine (Fig. 2.8). In this type of regulator, for one warp beam (Fig. 2.8(a)), the driving moment M_m created by the warp beam tends to turn gear wheels Z_5, Z_3 and worm Z_2, driving the warp beam. These movements are resisted by the worm Z_1: the teeth of the gear wheel Z_2, led from a tension K by force K_{ct}, nestle on the teeth of the worm creating counteracting force T_{ct} (at point C_t). Unlocking the gear wheel Z_2 with worm Z_1 is achieved with the shaft (2), giving constant rotation through a shaft (3) to a frictional disk (4). As soon as the roller (6) rests against the linear cam block (5) on rotation, the shaft (3) will be displaced to the right and close frictional clutch elements (4, 7). The clutch, by means of its spline coupling transfers rotation to worm Z_1. Force T_{ct} ceases to counteract force K_{ct} and the warp tension K is then able to turn the warp beam and gear wheels Z_2, Z_3 and Z_5. This will initiate a release of warp into the weaving machine. The amount of release is defined by the duration of contact with the frictional clutch (7) which, in turn, depends on the position of a roller (6) in relation to the linear cam block (5).

The regulator contains only the warp tension gauge, which is the back-rest (8). There is no warp beam diameter gauge. Therefore, the correction of the amount of warp released is made only according to the deviation of the back rest (8) over the set range of fluctuations. The responsive mechanism of the back-rest, as on all other weaving machines, is counterbalanced by springs F. In case of an increase in tension K, the back-rest (8) starts to fluctuate in a lower position. Thus, the shoulder (9) of the responsive mechanism of the back-rest moves the shoulder (12) upwards by means of a link (11) and the link lever (13) falls. The stationary pin (14) pushes the shoulder (15) to the left (a to a_1, Fig. 2.8), which applies the roller (6) to cam (5). The duration of contact of the linear cam (5) and the roller (6) increases and, hence, the duration of engagement of the frictional clutch (4, 7) also increases. The worm Z_1 turns over a greater angle and permits the rotation of the gear wheels Z_2, Z_3, Z_5 and the warp beam by tension K accordingly. Following the release of the warp, the tension of threads K decreases to the normal level.

The absence of a warp beam diameter gauge causes a given regulator to work in an *undesirable* mode that gradually *increases* the warp tension: a constant amount of warp release of the warp as the warp diameter diminishes is only possible through the fluctuation of the back-rest in a lower position, as found on belt (friction) regulators and hydraulic regulators. Tension K_s of the ESFF is regulated by the tension of springs F, the length of the shoulder ℓ_1, and the initial position of the roller (6) in relation to the linear cam (5).

Figure 2.8(c) and (d) present a diagram of a differential for the alignment of warp tension from two warp beams on one weaving machine. Driving moment M_{m1} (c) of warp beam *No. 1* turns gear wheels Z_5, Z_3, the sun wheel of differential a, and satellite c. The turning moment $M_{m2}=K_2^*p_2$ of warp beam *No. 2* turns gear wheels Z_6, Z_4, sunwheel b and satellite d. As a result, in position A (d) of the gearing of satellites c and d, two counteracting forces K_{11} and K_{21} result from warp tensions K_1 and K_2. If moments M_{m1} and M_{m2} are equal, then forces K_{11} and K_{21} will be equal. The satellites will remain motionless in relation to their axes, but they will rotate together with sun gears a and b, as the worm gear wheel Z_2 is turned by the worm Z_1. If tension K_1 exceeds K_2, and $M_{m1} > M_{m2}$, then at the point A, force $K_{11} > K_{21}$.

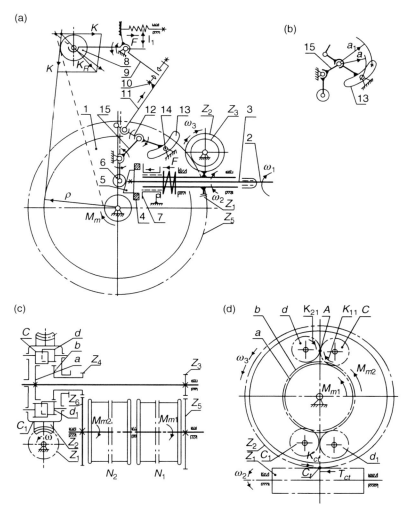

2.8 Differential frictional warp regulator (Sulzer). (a) Diagram of the connection of a warp beam with a worm mechanism and back-rest; (b) link lever; (c), (d) differential.

In this case, satellite *c* will force the rotation of satellite *d* in the *opposite direction*, which will turn sun gear *b*, gear wheels Z_4, Z_6 and *warp beam* No. 2 in the *opposite direction*. The tension K_{21} of the warp on warp beam No. 2 will rise, and the equality of moments M_{m1} and M_{m2} will be restored. The satellites will, once more, cease to rotate about the axes. Thus, the differential observes the equality of driving moments M_{m1} and M_{m2} from the two warp beams. Hence, a very important conclusion follows: at $\rho_1 = \rho_2$, the differential will support **equality of tension** of warp threads $K_1 = K_2$; at $p_1 > p_2$ the differential will support the inequality of tension ($K_1 < K_2$).

In practice, weavers have a major difficulty when forming a single wide length of woven fabric from TWO warp beams. The difference of tension of warp threads may increase gradually over time during the unwinding of the two warp beams. As a result, one half of the fabric (in width) will be formed at a higher tension of warp threads, leading to a more stretch-deformed condition of threads in the woven fabric structure. At the completion of weaving and on removal of the fabric from the weaving machine, the half of the fabric woven under the lower tension will become slacker and wavy. It is therefore necessary to establish identical *diameters* on the warp beams during weaving.

2.4 Condition of the equilibrium of the mechanism of a moving back-rest

The quality of formation and uniformity of the structure of a woven fabric depends on the stabilization of fluctuation of the back-rest. An understanding of the forces acting on the back-rest is of practical interest. Figure 2.9 presents a diagram of the typical behaviour of warp thread tensions K_1 and K_2, forces of springs (1) (2F) and gravity forces G of the back-rest (2) and levers (3) (the example is illustrated on a Sulzer machine).

The conditions of equilibrium of the forces affecting the back-rest is given by the equation of the sum of moments of all forces relative to axis **O**:

$$K_1 \cdot \ell_1 - K_2 \cdot \ell_2 + G \cdot \ell_3 - 2F \cdot \ell_4 = 0, \qquad [2.8]$$

where: K_1, K_2 = the warp threads tension;
$\ell_1 - \ell_4$ = the length of shoulders of corresponding forces: K_1, K_2, G, F;
G = the effective mass of the back-rest and levers;
F = the spring tension.

According to equality $K_1 = K_2$, we can define the spring tension F:

$$F = \frac{K_2(\ell_1 - \ell_2) + G \cdot \ell_3}{2\ell_4} \qquad [2.9]$$

where: K_1, K_2 = the warp threads tension;
$\ell_1 - \ell_4$ = the length of the shoulders of corresponding forces: K_1, K_2, G, F;
G = the effective mass of the back-rest and levers.

Mechanisms for warp release and warp tension control

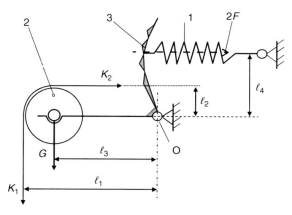

2.9 Action of forces of tension of warp threads, spring tension and gravity force of back-rest and levers.

For example, for $K_2 = 600$ N; $G = 400$ N; $\ell_1 = 217$ mm; $\ell_2 = 67$ mm; $\ell_3 = 150$ mm; $\ell_4 = 160$ mm;

we obtain: $F = \dfrac{600(217-67) + 400 \cdot 150}{2 \times 160} = 468.75\,\text{N}$

By knowing the value of force F of the tension, we can determine the primary tension setting of the warp threads K_2:

$$K_2 = \frac{2F \cdot \ell_4 - G \cdot \ell_3}{\ell_1 - \ell_2} \qquad [2.10]$$

where: $\ell_1 - \ell_4 =$ the length of the shoulders of corresponding forces: K_1, K_2, G, F;
$G =$ the effective weight of the back-rest and levers;
$F =$ the spring tension.

2.5 Stabilization of the mode of release and the tensioning of warp threads

The methods for constant warp release and warp tension maintenance on a weaving machine during the take-up of the warp beam is a challenge which has not been fully solved in the mechanisms considered so far. Manufacturability (i.e. the optimal functioning of the threads on the weaving machine) demands that the warp release mechanisms satisfy two conditions:

1. that the extent of linear release of the warp remains constant, so as to provide a constant tension in this component of the machine;
2. that the back-rest should fluctuate cyclically over a set range.

The first condition is not fully satisfied with brakes and regulators: at small diameters of warp beam, when the back-rest fluctuates in a lowered position, the set size $\Delta\ell$ is partially provided with a higher level of tension. Violation of the second condition leads to a change positioning the height of the back-rest position in relation to the neutral line of the shed and, as a consequence, to the change of the form (i.e. symmetry) of the shed and the conditions of formation of an element of the fabric. Understanding these problems allows for the devising of methods of stabilization of the release mode and the warp tension.

The simplest solution to this problem is carried out by introducing compensating mechanisms (in the form of profile plates, levers, clutches, etc.), the working elements of these mechanisms changing their position in the process of unwinding of the warp beam. However, the profile of compensating levers is effective only in relation to a given narrow range of fabrics. For the development of a fabric of any other structure, it will be necessary to calculate and design other compensator profiles. A further weakness in the majority of designs is the lack of communication between the tension level of the warp and the position of the compensator. Thus, compensating devices do not entirely resolve the problem of stabilizing the mode of release and the tension of the warp. In addition to the change in the driving moment of the warp beam M_m as the beam radius diminishes, it is necessary to consider the change of the angle of the warp sheet from the warp beam to the back rail, as this has an effect on the effective tension K_R.

This variation causes the initial adjustment of the tension setting of the ESFF to be disturbed: counterbalancing the back-rest moment due to warp yarn with the use of springs is still a current practice. However, the moment of action of effective tension K_R decreases. This difficulty can be eliminated simply by installing an additional back-rest between the warp beam and the main back-rest.

2.6 Comparative analysis of brakes and regulators

The *comparative analysis* of brakes and regulators highlights the contrasting areas of their application:

- In the formation densely woven fabric, more severe constraints of the warp release can be achieved by means of brakes.
- Warp regulators can provide a higher quality of uniformity of weft distribution in the fabric.
- For simple interlacings of the fabric, a mechanism providing a constant release of the warp at a constant set amplitude of cyclic fluctuation of threads tension regardless of the diameter of the warp beam by means of a movement sensor can be recognized as an advanced method.
- For the development of a fabric with a variable structure, a mechanism providing the variable release of the warp is desirable, while maintaining a constant warp tension irrespective of the warp beam diameter.

2.7 Questions for self-assessment

1. Why is the length of warp threads released from the warp beam longer than the length of the element of woven fabric produced from it?
2. What is the basic difference between the concepts of 'release' and 'feeding' of warp threads from the warp beam?
3. What function does the brake on a warp beam execute: the release or the feeding of warp threads from the warp beam?
4. What are the main forces which act on a warp beam on a weaving machine during its operation?
5. What happens to the warp threads tension on a simple band brake if the braking force is not regulated?
6. How does the Hartmann differential brake mechanism maintain constant warp tension?
7. At what position does the roller apply greater force on the differential lever in the Hartmann brake?
8. How does the automatic brake in the Elitex pneumatic weaving machine differ from the brake in the Hartmann design?
9. Why is the braking force on the warp beam on the Elitex weaving machine considerably less than that on the Hartmann weaving?
10. How is the automatic alignment of the tension of threads achieved on the Elitex (Kovo) pneumatic weaving machine?
11. How does a warp regulator differ from a warp brake?
12. What are the features that make the worm gear in a warp tension regulator independent of the tension of threads?
13. On a warp regulator, does a tooth of a worm wheel press on a tooth of a worm gear, or vice versa? Why?
14. For what purpose is the planetary mechanism in the Röper type of warp regulator used?

15. Can the planetary mechanism turn the warp beam in the Röper type of regulator?
16. Where do the two resultant forces from the tension of the warp threads and from the sley blade in the Röper regulator meet?
17. How is the automatic regulation of the amount of threads released from a warp beam obtained?
18. How it is possible to regulate a set level of threads tension on the Röper warp regulator?
19. Where is it necessary to place the back-rest for higher-count fabrics in the Röper regulator?
20. What are the main features of the Hunt warp regulator?
21. What are the deficiencies of the Hunt regulator?
22. Where do the two resultant forces from the tension of the warp threads and the variator drive meet in the Hunt regulator?
23. What are the features of a hydraulic warp regulator?
24. What is the sequence of actions in a Sulzer frictional warp regulator at warp release?
25. Why is it that, during the process of reduction of the diameter of the warp beam, the back-rest should fall gradually for the maintenance of a constant amount of threads release of from the warp beam on a Sulzer weaving machine?
26. For what purpose is the differential in the Sulzer warp regulator used? How does it work?
27. On which weaving machine are two resultant forces from two warp beams applied in the differential?
28. What supports the differential regulator of a Sulzer wide weaving machine: the equality of the tension of warp threads from two warp beams, or the equality of two turning points regarding warp beam axes?
29. How is it possible to maintain the primary warp tension level on a Sulzer regulator?
30. How is automatic alignment of two resultant forces from warp beams in the differential achieved?
31. What condition is necessary for the maintenance of a stable level of threads tension from two warp beams on the Sulzer weaving machine?
32. What condition of balance of the moments of the forces operates on the back-rest of the Sulzer weaving machine?
33. How it is possible to calculate the necessary spring tensions and warp threads tension on the Sulzer weaving machine?
34. What are the conditions of stabilization of the release mode and warp threads tension control on a weaving machine?

35. Carry out a comparative analysis of the use of brakes and regulators for the release of warp threads from the warp beam.
36. What are the more advanced methods of achieving perfection from warp regulators?

2.8 References

1. Gordeev V.A. and Volkov P.V., 'Weaving', 'Leg. and Pitsh. Prom.', Moskow, 1984 (in Russian).
2. Choogin V.V., Kahramanova L.F. and Nedovisiy M.N., 'Technology of Weaving Manufacture', State Technical University, Kherson, 2008 (in Russian).

3
Warp shedding in weaving: parameters and devices

DOI: 10.1533/9780857097859.40

Abstract: This chapter describes the geometry of the three parts of the warp shed, extending from the fell of the woven fabric to the back-rest. All types of shed, phases of their formation and warp thread elongation caused by the movement of healds are examined, and ways of reducing warp thread elongation are identified. The basic types of shedding device, and dobby and Jacquard machines are also described.

Key words: geometry of the components of the warp shed, types of shed, shedding devices, dobbies, Jacquard machines.

3.1 Introduction: parameters of the shed

The shed formation process is carried out by raising or lowering the heald shafts (or heald frames), each of which carries a number of healds through which the weft threads are individually drawn according to the structure of the fabric to be woven (Ref. 1). The process of lifting and lowering of the heald shafts together with the warp threads is called 'shed formation', or simply, 'shedding'.

The main geometrical characteristics of the shed are shown in Fig. 3.1(a) (Ref. 2 and Ref. 3):

- A is the position of the cloth fell, as denoted by the last pick added to the fabric. Note that A is towards the front of the weaving machine.
- B is the mid-position (shown dotted) of the heald shaft.
- NL is the neutral line of the shed, connecting the cloth fell A and the centre of the eye of the heald B at the mid-position (shown dotted) of the heald shaft.
- H_i is the height of the shed formed by the i-th heals shaft, defined by the apex of the vertical displacement of the warp threads, which are drawn into the healds of the i-th harness.
- L is the length of the shed from the cloth fell to the front dropper bar C.

- ℓ_{51} is the length of the front part of the shed from the cloth fell to the first heald shaft.
- ℓ_{41} is the length of the bottom branch of the back part of the shed from the first heald shaft to the dropper bar.
- ℓ_{4ij} is the length of the back part of the top branch of the shed from the i-th heald shaft to the j-th line of the dropper bar.
- h_i is the deviation of the thread, which is drawn into a heddle of the i-th heald shaft from the mid-position of heald shaft travel (so that $H_i = 2h_i$, etc.).
- 2α is the shed angle of the front part of the shed, where α is the deviation angle of the thread from the neutral line (NL).
- $+\Delta h_c$ $(-\Delta h_c^1)$ is the deviation of the threads of the top (or bottom) branch of the shed in the dropper bar from the NL.

Depending on the method of shed formation, a full shed or a non-full shed (or half-shed) can be obtained. A full shed is obtained by raising particular warp threads to the top position and lowering the rest to their bottom position, as required by the weave (Fig. 3.1(a)). By raising some warp threads to their top position and leaving the rest in the NL position, a **top half-shed** is obtained. Similarly, it is possible to obtain a **bottom half-shed** by lowering some of the warp threads and leaving the remainder in the NL position (Fig. 3.1(c)).

By suitable adjustment of the amount of lifting or lowering movement given to different heald shafts, the front part of the shed (front shed) obtained can be called an 'even' (or 'clear') shed (Fig. 3.1(a)), or 'uneven' (or 'irregular') shed (Fig. 3.1(b)). When the shed is open, depending on the position of the warp threads in the front part of the shed, it can be even, uneven or mixed. When the amount of movement of heald shafts is similar (Fig. 3.1(d)), the threads in the front part of the shed are located in different planes, giving an *uneven* shed. By adjusting the amount of movement of each heald shaft appropriately, so that the movement of the heald shafts gradually increases towards the back, an *even* shed can be obtained (Fig. 3.1(a)). This ensures the warp threads in the front part of each shed line up in the same plane, which improves the picking of the weft thread in the shed. A *mixed* shed is obtained (Fig. 3.1(e)) when one of the shed lines (the upper or, often, the bottom shed line) is set up as an even shed, and the other is set up as an uneven shed. According to the characteristics of heald shaft movement in the shedding cycle, two types of shedding movement can be obtained: open shedding or closed shedding.

Open shedding (Fig. 3.1(f)) is obtained when those heald shafts which need to be in the up position or down position over a number of successive weft insertions in the shedding cycle are allowed to remain continuously in

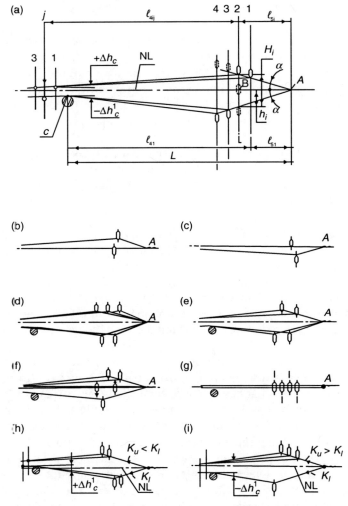

3.1 Types and parameters of shed. (a) The shed on a weaving machine; (b), (c) non-full shed; (d) uneven shed; (e) even shed; (f), (g) level phase of the warp threads; (h), (i) variable tension of shed. Refer to text for detailed explanation of geometrical characteristics.

the up position or in the down position, as appropriate, during that period. These heald shafts do not move to the middle line of the shed, but stand in the up or the down position over a certain number of weaving cycles, as permitted by the fabric structure. Only those heald shafts that must change their position move through the middle level during the same period. This reduces the amount of abrasion to these warp threads due to warp cross over, and improves the conditions of the weft insertion in the shed. The reduced shed movements involved in open shedding also save some energy (compared with closed shedding, as will be seen). However, the tension of the warp threads drawn in standing heald shafts is uneven compared with that in the heald shafts that move during the same period, and this has a negative influence on the beating-up of the weft and, hence, on the fabric structure.

When *closed* shedding is used (Fig. 3.1(g)), every heald shaft moves to its centre position in each weaving machine cycle before it moves to its open position in the next cycle. With this type of shedding movement, the warp threads change their tension equally during the process of shedding, and this helps to maintain the even allocation of the load between the warp threads during the fabric formation, which improves the operation of the weaving machine. However, some increased attrition of the warp threads is likely to occur because all threads move in a congested manner during shedding. It is better to use open shedding for cotton, linen, silk and woollen yarn during the formation of fabrics with a low density and many repeats of interlacing. Closed shedding is normally used for fabrics from yarns which have a high level of unevenness, for example, carded woollen yarns.

To facilitate the process of fabric formation, a **variable tension** (or **unbalanced**) **shed** geometry can be employed. By raising the dropper bar in relation to NL by $+\Delta h_c$, the bottom (lower) branch of the shed is stretched more (Fig. 3.1(h)) so that $K_\ell > K_u$, and when the shed is opened, the bottom shed branch will have a higher tension than the top branch. This is called '**positive variable tension shedding**'. In contrast, by lowering the dropper bar by $-h_c^1$, in relation to NL (Fig. 3.1(i)), and when the shed is opened, the top (upper) branch of the shed is stretched more than the bottom one and $K_u > K_\ell$, we have **negative** variable tension of the shed branches. An equally tensioned (**balanced**) shed is presented in Fig. 3.1(a). The upper surface of the back-rest (2) (See Chapter 1, Fig. 1.1) is placed along the height of the weaving machine along the bisector of the shed on the part ℓ_3 (from the back-rest) to the dropper (Fig. 1.1).

When correctly set, the droppers should have sufficient vertical movement. If the movement is too small, especially that of the back droppers (those most distant from the harness), this can cause the jamming of knots and the accumulation of fluff in front of the droppers, as well as thread breakages. The duration of the shed staying in a fixed position, or the duration of movement of the warp threads over a specified part of the shedding (usually expressed by the corresponding angle of rotation of the main shaft) is known as the 'dwell period'.

The **level phase** of the warp threads is the angle (φ_1) of rotation of the main shaft during the period when the warp *threads are briefly motionless (since the heald eyes are longer than the diameter of warp threads)* on the middle level of the shed while the *heald movement* is in progress (Fig. 3.2(I)). On average, $\varphi_1 = 20°, \ldots, 25°$.

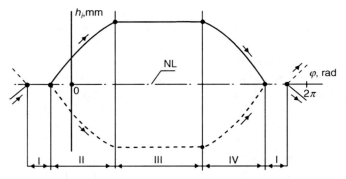

3.2 Phases of shedding of the warp threads.

The **shed opening phase** is when the warp threads are passed from the middle level to the full opening of the shed (Fig. 3.2(II)), $\varphi_2 = 90°, \ldots, 130°$.

The **dwelling phase** is the length of time warp threads stay in the position of full opening of the shed, φ_3 (Fig. 3.2(III)), $\varphi_3 = \varphi_2 + (90°, \ldots, 120°)$.

The **shed closing phase** is the length of time required to shift warp threads from the dwelling position to the level position φ_4 (Fig. 3.2(IV)). As a rule $\varphi_4 = \varphi_2$.

The four aspects of thread levelling are: duration of levelling, moment of levelling, heald shaft levelling and the size (magnitude) of levelling. The duration of levelling depends on the size of the heald eyes and on their traverse speed. The moment of levelling is the moment of shed formation, when the healds begin to move (displace) the threads from their resting position in the level phase. Heald shaft levelling means the alignment of heald frames by an opposing motion in the thread levelling phase. Size of levelling will be considered in Section 6.3 (pp. 114–116).

3.2 Elongation of warp threads in shedding

For clean shed formation, it is necessary to increase h_i, the displacement of threads by the healds being proportional to their distance from the cloth fell.

The ratios, $h_1 : h_2 : h_3 \cdots = \ell_{51} : \ell_{52} : \ell_{53} \cdots$. Different heights of thread deflection, as a result of drawing in through the different heald shafts, lead to a difference in thread elongation. For the i-th heald shaft, the total (front and back part of the shed) thread elongation due to shedding in the top branch of the shed according to Fig. 3.1(a) will be given by the expression (Ref. 2 and Ref. 3):

$$\lambda_{hij} = \lambda_{5i} + \lambda_{4ij} = 0.5 \cdot \left[\frac{h_i^2}{\ell_{5i}} + \frac{(h_i - \Delta h_c)^2}{\ell_{4ij}} \right] \quad [3.1]$$

Warp shedding in weaving: parameters and devices   45

For example, for seven heald shafts ($i = 7$) in a Sulzer weaving machine $\ell_{47j} = 358$ mm; $\ell_{57} = 191$ mm; $h_7 = 40.59$ mm; $\Delta h_c = 6$ mm; we obtain:

$$\lambda_{h7j} = \lambda_{57} + \lambda_{47j} = 0.5 \cdot \left[\frac{40.59^2}{191} + \frac{(40.59-6)^2}{358} \right] = 5.98 \text{ mm}$$

Total thread elongation due to shedding in the bottom branch of the shed:

$$\lambda_{hi} = \lambda_{5i} + \lambda_{4i} = 0.5 \cdot \left[\frac{h_i^2}{\ell_{5i}} + \frac{(h_i - \Delta h_c^1)^2}{\ell_{4i}} \right] \quad [3.2]$$

For $\ell_{47} = 255$ mm and $\Delta h_c = \Delta h_c^1 = 6$ mm, we obtain:

$$\lambda_{h7j} = \lambda_{57} + \lambda_{47} = 0.5 \cdot \left[\frac{40.57^2}{191} + \frac{(40.57-6)^2}{255} \right] = 6.65 \text{ mm}$$

Practical experience has shown that equal tension exists in the front and back parts of the shed. From Equation [3.1] we can define the condition of *equal elongation* of parts of a warp thread in the front and back parts of the shed.

For the top branch:

$$\lambda_{5i} = \lambda_{4ij} = \frac{h_i^2}{2\ell_{5i}} = \frac{(h_i - \Delta h_c)^2}{2\ell_{4ij}} \quad [3.3]$$

For the bottom branch:

$$\lambda_{5i} = \lambda_{4i} = \frac{h_i^2}{2\ell_{5i}} = \frac{(h_i - \Delta h_c^1)^2}{2\ell_{4i}} \quad [3.4]$$

Thus, equal elongation is possible in this case:

$$\ell_{4ij} = \ell_{5i} \frac{(h_i - \Delta h_c)^2}{h_i^2} \quad \text{and} \quad \ell_{4i} = \ell_{5i} \frac{(h_i - \Delta h_c^1)^2}{h_i^2} \quad [3.5]$$

Equal elongation in the *front* and *back* parts of the shed (having indices by subscripts 5 and 4) of thread lead to *equal tensions* of these parts of thread. By inserting the actual values of ℓ_{5i} and ℓ_{4ij} or ℓ_{4i}, Δh_c and Δh_c^1 for *each* heald

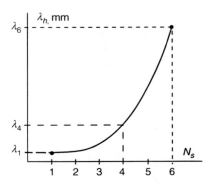

3.3 Variation of thread tension, according to heald shaft position.

shaft $i = 5, 6, 7, 8, 9, 10$ and 11 in Equations [3.3] and [3.4] (for a given type of weaving machine), we can determine the *number i* of the heald shaft which has equal values of elongations $\lambda_{4ij} = \lambda_{5i}$ and $\lambda_{4i} = \lambda_{5i}$ for the top and bottom branches. Note here, importantly, that ℓ_{4ij} is always more than ℓ_{4i} (see Fig. 3.1), therefore every thread at the top position of the heald shaft i *always* has a smaller value $\lambda_{hij} < \lambda_{hi}$ as compared with the bottom position of the heald shaft.

Figure 3.3 shows how λ_h increases in the heald shaft as number N_s increases, so it is sensible to try to produce the fabric using the minimum number of heald shafts.

3.2.1 The cyclogram of shedding

Since the elongation λ_h of warp threads depends on the heald shaft location relative to the fabric fell and droppers, the variation in λ_h depends on the type of shed and the nature of the shedding cycle for each warp thread in the weave repeat (Ref. 2). Figure 3.4(a) shows the weave repeat of a complex twill 2/2 + 1/1, and also the cyclogram of variation of λ_h for the first warp thread in a weave repeat for closed shedding (Fig. 3.4(b)) and open shedding (Fig. 3.4(c)). The other threads in the weave repeat elongate identically but with a shift of 60°.

Cyclogram analysis of different weave interlacings shows that it is necessary to identify the heald types *separately* for the top and bottom branches of the shed (upper and lower parts); the type of weave interlacing (shedding cycle formula); and the type of shedding mechanism used.

For example, when plain weave (shedding cycle 1/1) is produced, $0.5n_{wp}$ warp threads (where n_{wp} = the total quantity of warp threads in woven fabric) go up and the other $0.5n_{wp}$ warp threads go down across the neutral line

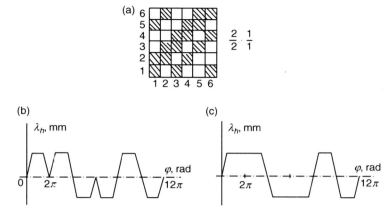

3.4 The shedding cycle. (a) Interlacing (weave) repeat; (b) closed shed; (c) open shed.

NL at the insertion of every weft thread. As a result, all the warp threads are involved in the level phase of shedding. Above and below this (NL), no warp threads remain standing in the open-shed position.

When twill with single warp overlapping 1/3 (or 1/4, ...) is produced, $0.25n_{wp}$ warp threads go up and $0.25n_{wp}$ warp threads go down to the neutral line (NL). This group of threads stops in the level phase. However, $0.5n_{wp}$ warp threads remain in the down position (the bottom branch of the shed); hence, as the threads form the next position of the level phase, there are threads below the NL, but none above it.

When twill with long warp overlaps is produced, for example, 2/3 (or 3/4, 3/5, ...), $0.20n_{wp}$ warp threads remain in the up position and $0.40n_{wp}$ warp threads remain in the bottom branch of the shed in the level phase, $0.20n_{wp}$ warp threads move to the NL from the top branch, and $0.20n_{wp}$ warp threads move to the NL from the bottom branch. The top and bottom branches are therefore always in level phase.

It should be emphasized that this arrangement, which only applies to an open shedding type dobby, must be used to calculate the tension of every warp thread over an interlacing repeat in the beating-up phase of every weft thread into the cloth fell. For example, plain weave (1/1) is produced only by *closed* shedding, regardless of the type of shedding device. Twill with single warp overlapping (1/2, 1/3, etc.) is produced with the top branch of the shed according to the neutral line only by *closed* shedding, and the bottom branch of the shed can be produced by closed *or* open shedding, subject to the construction of the shed. Twill with long warp overlaps (floats) (e.g. 2/2,

2/3, 3/4, etc.) can be produced by closed or open shedding, or a combination of the two.

3.2.2 Methods of reducing yarn deformation due to thread tension in shedding

The lowest level of strain λ_h and, consequently, a safe minimum level of warp thread tension needs to be maintained in the warp during the levelling phase, to avoid the droppers 'sagging' or 'dropping' and also to avoid flawed interlacing due to warp threads sticking when the shed is opened (Ref. 1 and Ref. 2).

This is most widely achieved by reducing warp thread elongation (strain) with the help of a rocking beam (Keighley, Fig. 3.5(a)). At the opening of the shed, the back-rest (3) moves towards the heald shafts, reducing the warp strain λ_h and, hence, the thread tension K arising at the warp beam (2). The backward motion of the back-rest and the restoration of the warp thread tension happen at the closing of the shed. The back-rest (3) is driven by the eccentric cam (1) by means of a double-armed lever (4). A rocking back-rest should be used in closed shedding weaving.

The leasing device (Fig. 3.5(b)) is also widely used. It consists of dividing rods (5, 6), which are rocked by the triple-armed lever (4) of the eccentric cam (1). The figure shows the lowest position of the rods (5, 6), where the top branch of the shed is stretched more than the bottom one. When the rods are at the same level on the line NL, both branches of the shed will have the same tension (by stretch deformation). The bottom branch of the shed will be strained more if the upper position of rods (5,6) is limited. Thus, the difference in tension of the shed can be controlled by adjusting the rocking angle of the lever (4).

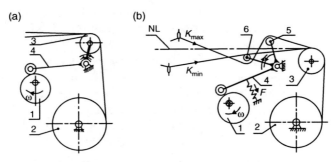

3.5 Mechanisms of reducing of thread deformation under shedding. (a) Rocking beam (Keighley); (b) leasing device.

3.3 The classification of shedding devices

There are two main groups of mechanisms (Ref. 1 and Ref. 2):

1. Tappet mechanisms, which produce both heald shaft lifting and lowering actions with one working component – the tappet (also called the 'cam').
2. Dobbies and Jacquard machines, where each function (the formation of the shed and the control of the shed changing process) is carried out by a specific mechanism.

Tappet mechanisms, with dependent (negative) or independent (positive) harness motion, where the heald shaft drive mechanism is located either inside or outside the weaving machine frame and has either flexible or rigid connections, belong to the first group.

Dobbies can be of the single-lift or the double-lift type, with, respectively, one or two rotations of the main shaft providing the entire cycle of the moving parts; they can provide closed- or open-shed movement; and they may be single-shaft or twin-shaft (according to the number of 'prisms' carrying punched cards).

Jacquard machines are classified according to the distance between the centres of holes A on the punched cards which carry weave information:

Long – A = 6.83 or 6.50 mm
Medium – A = 5.75 or 4.64 mm
Short – A = 4.30 or 3.99 and 2.85 mm
Shortest – A = 2.66 mm.

In addition, Jacquard machines are classified by power according to the number of hooks used, ranging from 100 to 2600.

3.4 Tappet shedding

The first mechanical device to be used for shedding movements in weaving was the mechanical tappet (cam) gear. Its simplicity, multifunctionality and reliability contributed to the widespread use of cam shedding, and it remains in use in modern weaving machines (Ref. 1 and Ref. 2). On the Hogdson shuttle weaving machine, a cam mechanism with dependent (or negative) motion of the heald shafts is used for shedding in the production of the simplest type of fabrics. In this drive (Fig. 3.6), the heald shafts (1, 2) are connected by flexible bands (3, 4) and treadle levers (5, 6) with their rollers pressed onto the tappets (profiled cams) (8, 9) under tension

50   Mechanisms of flat weaving technology

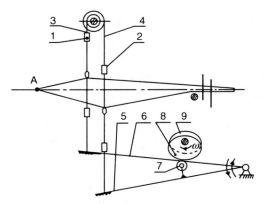

3.6 Tappet gear (Hogdson), with the related movements of the heald shafts with flexible heald ties in the shuttle weaving machine.

of heald ties (3, 4). At higher densities of warp threads, the two heald shafts are joined to every heald tie (heald shaft cord). For fabric of other interlacings, the heald shafts should have independent movements. Uncontrolled motion is avoided by force closure of cams by spring action on the treadle levers. This prevents the shuttle weaving machine speed from rising above 240 picks/min.

Positive tappets are used to drive the heald shafts on faster weaving machines, such as the air-jet or water-jet types, and directly control raising and lowering of the heald shafts. These tappets may be of the grooved disc (Elitex) or double disc (Sulzer) type. Figure 3.7 shows the drive gear for the Elitex pneumatic weaving machine harness. Independent movement of the heald shaft (1) is carried out by the grooved tappet (7) with the help of rigid connecting rods (constraints) (2, 3, 4, 5, 6). The fabric structure can contain up to 10 various interlacings of warp threads. Each heald shaft requires its own tappet. A weaving machine with this type of tappet drive can operate at a speed of over 400 min^{-1} (rpm) on the main shaft. Tappets can be used for the creation of interlacings with a shedding cycle such as 1/1; 2/1; 2/2; 1/3; 1/4, etc. If the connecting rod (4) is connected in the alternative position (shown by a dashed line), instead of a cycle 1/3, a cycle 3/1 is obtained. The disadvantage of these drives is their open construction under the heald shaft, which tends to cause the harness grooves to become blocked with lint.

Figure 3.8 shows the design of possibly the most widely used variant of dual-tappet heald shaft drive, which is used in the Sulzer weaving machine. The structure of the tappet and counter-tappet arrangement (1) allows accurate and close force closure by three-arm lever (2) using rollers (3). The term 'force closure' means that the two elements concerned, the tappets and the three-arm lever, are not free to move uncontrollably, as contact between

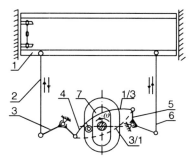

3.7 Independent movement of the heald shafts with a tappet gear with rigid connecting rods on a pneumatic weaving machine (Elitex).

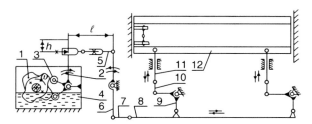

3.8 Independent movement of a heald shaft with a positive tappet drive (Sulzer).

the tappet discs and the lever (2) is always maintained. The position of the gear in a closed oil reservoir (4) provides for reliable and durable operation of the tappets. The motion of the related heald shaft is carried out by rigid connecting rods (constraints) (5, 6, 7, 8, 9, 10). Each heald shaft is controlled by one pair of tappets. Experience shows that a tappet drive based on 10 heald shafts can provide for the production of 70% of high-volume fabric types. This arrangement makes possible the easy control of the angle 2α of shed opening (shed height) from 16° to 26° by changing the size h, and also the shed position or height relative to the sley of the weaving machine (shed timing) with the length ℓ of link (5). In general, this type of heald shaft drive does not restrict the speed of the weaving machine.

The basic disadvantage of tappet shedding is that each different shedding sequence requires a particular type of tappet, which limits the flexibility of being able to produce different weaves for a given weaving enterprise. The essential advantage of tappets is that the profile of tappets and connections with a heald shaft defines *all parameters* of shedding: pattern of thread interlacing; heald shaft and thread order; duration of the dwell phases; duration of heald shaft motion; type of shed; cyclogram of warp thread motion; and height of shed.

As a rule, the nose of the tappet corresponds to the dwell of the heald shaft in the up position, and the hollow (or valley) of the tappet to the down (lowered) position. The speed of the tappet rotation is R_u times less than the speed of the main shaft rotation, depending on the weave appropriate to the tappet. The rotation R of the shedding device tappet is the number of rotations of the main shaft for *one* rotation of the tappet. The rotation can be equal to one repeat R_{wft} of the weave interlacing, $1R_{wft}$, $2R_{wft}$ or $3R_{wft}$, etc., that is, a multiple of the repeat of interlacing of the weft threads. For example, if the minimum rotation R of tappet shedding which was considered above on a weaving machine is equal to 4, it means that for manufacturing linen with 1/1 cycle, the rotation $R = 2R_{wft} = 4$, and for twill 1/3 $R = 1R_{wft} = 4$. On a weaving machine, the maximum rotation R of tappets is equal to 8. Any further increase of rotation would have a substantial adverse effect on the dynamic characteristics of the tappet.

Figure 3.9 shows a variety of tappet constructions with which to obtain the indicated number of shedding cycles (Ref. 2). The 1/1 plain weave fabric (Fig. 3.9(a)) can be woven using a two-rotation (two rotations of the weaving machine main shaft for one rotation of tappet) tappet (b) on the shuttle weaving machines, or a four-rotation tappet (c) on shuttleless weaving machines. The twill 1/2 (d) can be manufactured using a tappet with a rotation period $R = 3$ (e) on the shuttle weaving machines, or $R = 6$ on the

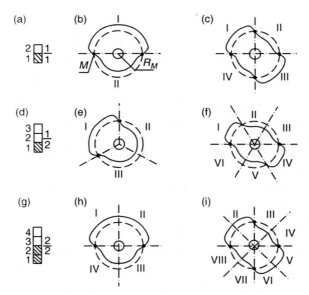

3.9 Construction of the sections of a tappet.(a), (b), (c) For linen 1/1; (d), (e), (f) for the 1/2 shedding cycle; (g), (h), (i) for the 2/2 cycle.

shuttleless weaving machines. The 2/2 twill (g) can be manufactured by a four-rotation tappet (h), or by an eight-rotation tappet (i). The tappet profiles for plain weave 1/1 and twill 2/2 (b and h, c and i) can be determined by comparing the cross-sections. Thus, it is possible to change the shedding cyclic diagram by varying the rotation period R of these tappets. For example, it is possible to manufacture the 2/6 twill using the four-rotation tappet ($R = 4$) instead of 1/3, reducing the number of tappet rotations by half ($R = 8$).

The following general shedding contour construction *algorithm* is recommended:

- Divide the plane with sectors (I, II, ...) with respect to the centre of the tappet disc according to the tappet rotation period required.
- Draw a dashed circle with average radius R_M, indicating the average heald shaft level (i.e. position when the heald shafts are level).
- Position the tappet noses (peaks) and hollows (valleys) at an equal distance from the dashed circle with arc segments using the shedding cycle formula.
- Connect the peaks and valleys with arc segments on the transitional areas via the levelling points M.

The tappet pack setting on the main shaft is carried out according to the weave repeat required in the fabric to be woven. For example, for normal drawing of warp threads in the heald shafts for the 1/3 twill weave shown (Fig. 3.10(a) and (b)), a four-rotation tappet contour is required. So, for this weave, interlacing warp thread (1) (warp threads are represented vertically) requires tappet 1 to lift the heald shaft through which warp thread (2) is drawn. For 1/3 twill interlacing, the tappet for warp thread (2) should be moved forward by $360°/4 = +90°$. The same consideration applies to warp threads (3) and (4).

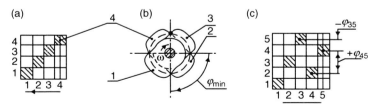

3.10 Tappet set. (a) One-and-three twill; (b) to alternate the tappets; (c) sateen 5/2.

In general, the minimum tappet shift angle is determined by the rotation period:

$\varphi_{min} = 360°/R_{wft}$, where R_{wft} is the number of weft threads in the weave repeat, whereas the shift angle φ_i depends on the overlap disposition in the interlacing repeat. Single warp overlaps are usually considered the basic ones. A variable shift angle φ_i is visually demonstrated by the 5-heald shaft movements of the 5/2 sateen (Fig. 3.10(c)): for which φ_{min} = 360°/5 = 72°. The tappet asperity shift for the 4th warp thread relative to the tappet asperity for the 5th thread: $\varphi_{45} = 72° \times 2 = +144°$. Thereafter, there is a negative shift $\varphi_{35} = -72°$; $\varphi_{25} = +72°$; $\varphi_{15} = +72° \times 3 = +216°$.

3.5 Dobbies

A wider range of fabrics can be produced with the help of dobbies providing control of the movement of 12, …, 14 or more heald shafts (Ref. 1 and Ref. 2). Depending on their design, dobbies can move through a full cycle in one rotation (single-lift type) or two rotations (double-lift type) of the weaving machine main shaft. The lower speed of double-lift dobbies makes them quite suitable for use on high-speed weaving machines. Modern high-speed dobbies are capable of making open or centre-closed sheds. Dobbies that only make centre sheds, such as Textima, are used only on shuttle weaving machines.

Many different types of dobbies have been created during 150 years of weaving machine development. All have mechanisms for the drive, the transfer of motion to operating parts and the control of the lifting order of the heald shafts, their dwelling and lowering. Modern versions of dobby are equipped with electronically programmed mechanisms instead of being fully mechanical devices, enabling woven fabric patterns to be changed quickly and simply. In addition, advanced models are electronically programmed to control a 2–8 colour weft selection device, and to create commercial inscriptions on the edge of the cloth. The main obstacle to the widespread introduction of electronically controlled dobbies is their high cost, which is commensurate with weaving machine prices.

The single-lift dobby design on 12 heald shafts of the open-shed type with flexible connections, as in the Knowles dobby, which is widely used on shuttle weaving machines for producing woollen fabric, is shown in Fig. 3.11(a). The dobby is driven from the main shaft (1) by means of gear wheels Z_1 to Z_4. Rotary motion is transferred by the shaft (2) to a two-segment toothed cylinder (3 and 4), resembling wheels (pinion gears) with teeth removed on half of the surface. If the prism (5) after rotary motion substitutes the small-diameter bush (6), the lever (7) will move down. Thus, the wheel

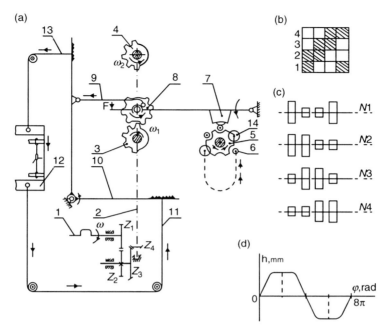

3.11 Single-lift dobby (Knowles). (a) Dobby mechanism; (b) twill repeat 2/2; (c) roller card; (d) harness frames movement cycle chart.

(pinion gear) (8) interlocks with the teeth of the lower rotating segment of the toothed cylinder (3) and the rotary motion relative to its axis will move the link (9) to the left, and together with it the double-arm lever (10), which will lower the heald shaft (12) by its horizontal lever arm by means of flexible cord (couplings) (11). The vertical lever arm (10) and flexible cord (coupling) (13) will lift the heald shaft (12) when the larger-diameter roller (14) is under the lever (7). The roller card (Fig. 3.11(c)) can easily be set up in accordance with the interlacing repeat (Fig. 3.11(b)) at a straight draft of cords in the heald shaft.

Heald shafts can remain in the up (lifted-up) or down (lower) position (Fig. 3.11(d)) for twill 2/2. Sharp periodic movement of the driven parts and the heald shafts helps to separate rough woollen yarns at the formation of the shed. However, impact loads where the toothed cylinders (porcupines) (3, 4) interlock with the tooth (8) lead to rapid wear. The dobby limits the speed of a wide weaving machine to about 180 min^{-1} (rpm).

For the production of cotton fabrics on shuttle weaving machines semi-open-shed double-lift Hattersley dobbies with flexible couplings have generally been used. This type of dobby lifts the heald shafts, which are lowered by return springs. Therefore, the heald shafts can remain stationary only in the down position, and when raised to the up position, they descend at the shed changeover (hence the name 'half-open-shed'). The double-lift design of the dobby enables it to be used on narrow, relatively high-speed machines at a speed of 220 min^{-1} (rpm).

In Fig. 3.12 (a), the design of the high-speed knife open-shed double-lift Staubli dobby with 14 heald shaft capacity is shown. The dobby contains 14 top main knives (1), and 14 low knives (2), held in place by main lifting hooks (3 and 4) with the balance lever (5), which transmits motion through linkages (6, 7, 8, 9, 10, 11, 12, 13) to the heald shaft (14). The main knives (1, 2) are designed for the forced lifting of heald shafts by means of a pantograph (Fig. 3.12(b)) from the grooved tappet (15). The heald shafts are lowered by push knives (16, 17), which in turn press on hinges (18, 19) and lead them to positive stops (20, 21). The main lifting hooks (3, 4) are operated

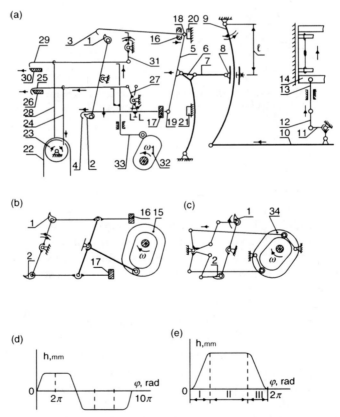

3.12 Staubli high-speed knife open-shed double-lift dobby. (a) Dobby mechanism; (b) main knives gear; (c) main knives gear rotary device; (d) 2/3 shedding cycle; (e) phases of heald shaft movements.

by means of perforated paperboard tape (22) carried on a prism (23). The needle (24) falls through the hole in the perforated tape, lowering the small lifting hook (25) on the small knife (26), the motion from which is transmitted to the low stop (27) to release the main lifting hook (4) into the path of the main knife (2). The heald shaft, which is connected with this main lifting hook, then ascends.

Where there is no hole in the paperboard for the needle (28), then the top small lifting hook (29) remains above the path of the small knife (30). The down stop (31) continues holding the top lifting hook (3) in the raised position over the path of the knife (1). The tappet (32) lifts and lowers the small lifting hooks with push bars (33). The main axle drive (Fig. 3.12(c)) consists of grooved tappet (34) and linkage. Angle 2α is regulated by the size ℓ (Fig. 3.12(a)) and clamp position (8) on the treadle lever (9).

Two rows of holes are assigned to each heald shaft on the punched paperboard tape (22). The prism turns once after two rotations of the machine's main shaft. The perforations on the tape are arranged according to the filling design pattern. The dobby works stably at a speed of 250 min^{-1} (rpm). In Fig. 3.12(d), the heald shaft movement for shedding cycle 2/3 is shown. Figure 3.12(e) shows the stage of motion dwell II of the heald shaft $< \varphi_2 = 120°$.

Figure 3.13 illustrates the assembly of the open-shed Staubli rotary dobby with 14 heald shafts for Sulzer weaving machines. This is an add-on device to the common tappet drive of harnesses on the Sulzer machine. A single gear wheel (2) is added, which meshes with the wheel (3) that is fixed on the splined shaft (4) on each pair of shedding tappets (1). In general, there are between 3 and 14 wheels. The heald shafts (5) are moved by means of the programming mechanism (6), which usually contains a punched tape, needles and spline transmitter. The shifting of the splines (7) alternates the use of the wheels (3 and 2). Therefore, the pattern of the holes on the punched tape does not directly correspond to the paperboard of the design pattern. The speed of machines with this type of dobby is 250 min^{-1} (rpm).

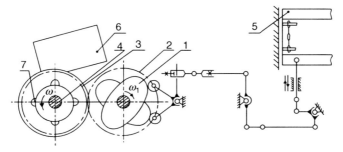

3.13 Open-shed rotary dobby (Staubli).

3.6 The Jacquard machine

J.M. Jacquard weaving machines are used for the management of small groups and even separate groups of warp threads during shed creation (Ref. 1 and Ref. 2). Individual healds (Fig. 3.14) are used, as individual warp thread control is required in Jacquard weaving. A Jacquard machine is installed over a weaving machine at a height of about 3 m to enable a reduction of the curvature of the special harness cord (5, 6) to which healds (3, 4) are attached. The Jacquard machine only lifts warp threads (1, 2) and they are lowered under gravity by lingos (24) or rubber tension bars. The thread lift is controlled, as on a dobby, by a special prism (22) and pasteboard punched cards (21).

Almost all single-lift and Jacquard machines are used in various configurations for the creation of full, short, centre-closed and open sheds. However, the basic operating units of the Jacquard machine all perform the same functions. Figure 3.14 shows the threading arrangement of a centre-closed Jacquard machine for the creation of a full single-lift shed with a 1300 hook capacity plus 16–48 hooks provision for the selvedges of the fabric.

Warp threads 1, 2 (Fig. 3.14(b)) are passed through the openings of healds (3, 4), which are connected with the harness cords (5, 6) (Fig. 3.14(a) and (b)). Every harness cord is passed through an opening in the comber board

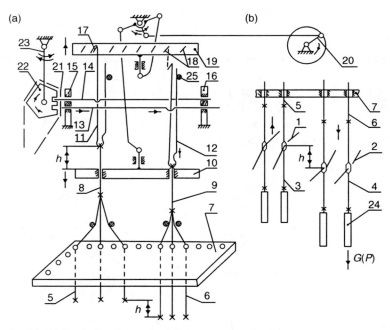

3.14 (a,b) Basic configuration of the Jacquard machine.

(7). A group of harness cords is attached to each of the neck cords (8, 9), which are passed through the opening of the trap (or collar) board (10). Every neck cord is attached to a heel hook (11, 12). The hooks are controlled by needle knees (13, 14) situated in the openings of needle boards (15, 16). In the level phase, all hooks are situated at one level below the knives (17, 18) of the griffe block (lifting bar) (19). The healds and threads are also situated at one middle (level) grade (shed closed). The griffe block (lifting bar) (19) and trap (levelling) board (10) take up a maximum close position. The board (10) and frame (19) are driven by a system of chains from a crank (20).

When it reaches the opening in the card (21), by moving the prism (22) to the right on the winch, the needle (14) goes through the opening in the prism and then remains immobile. The knife (17) then captures and lifts the hook (11), neck cord (8), harness cord (5), heddle (3) and load (24) (Fig. 3.14(b)), forming the upper part of the shed. At the same time, the section of card (21) that is without a hole pushes on the point of the needle (13) and displaces the hook (12) from the path of the knife (18) with the help of the crank (23). The knife (18) goes up past the hook (12). As the griffe block (lifting bar) (19) is raised, the trap board (10) is lowered with leaning hooks (12), neck cords (9), harness cord (6) and heddle (4). This forms the low branch of the shed. The elastic of the hook branches (12) presses on props (25) to return the needles (13) to their original position on the left. An alternative style of hooks without a long tiller is used for the springs for each needle in the area of the needle board (16). When the batten (23) is opened, the prism(22) rotates and presents a new card to control the machine needles. The number of needles is equal to the number of hooks. An aperture (hole) on the card means the lifting of the connected threads. The speed of a weaving machine with Jacquards is limited to between 120 and 150 min^{-1}.

In practice, two methods of tying up (or 'cleaning') the machine are used: (i) open with the harness cord in layers parallel to the reed; and (ii) crossed without harness cords in layers. When the open method is used, the prism with the cardboard is located in front of or behind the weaving machine; although this causes servicing problems, there are fewer harness cord scuffing problems.. The most widely used method is the crossed tie-up, where the cardboard is located on the side of the weaving machine.

The principal defect of the Jacquard machine is the force exerted by the points of the metal needles on the cardboard tabulating cards, causing rapid wear and frequent damage, and a need for replacement cards and the expense of stamping new apertures corresponding to the figure being woven. This defect is overcome by the Verdol additional needle device from France (Fig. 3.15(a)), which is located between the prism (1) and machine needles (3) (Fig. 3.15(b)).

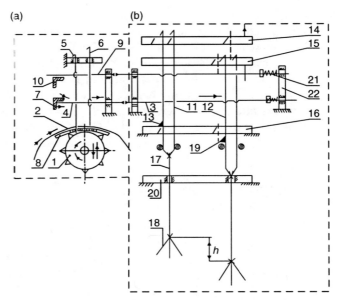

3.15 Double-lift open-shed (Jacquard). (a) Verdol; (b) Zangs.

This system works by the lightweight metal needles (4), and not the cardboard, bearing on the machine needles (3) (Fig. 3.15(b)). Only the thin, light vertical needles (5, 6) of the device are in contact with the cardboard, reducing card wear. The reciprocal motion of the knives (7) applies pressure to the needles (3) and vertically to the needles (4). If the continuous card (2) is in the form of a punched tape, then the raising of the pegged cylinder (prism) (1) by the needle board (8) keeps the needle (5) in the lowered position, away from the card. The needle (9) that is located on the horizontal rack of the knife (10), does not press on the machine needle. Thus, the thread is raised in accordance with the aperture on the card. The Verdol device has allowed the Jacquard machines to increase their power to 2600 hooks.

The obstacle to high-speed operation of Jacquard weaving machines is the slow descent of the heddles under gravity. A partial solution to this problem was proposed by the German company Zangs: the use of rubber tension bars instead of plummets or lingos. Nowadays, the widespread use of shuttleless weaving machines has resulted in high-speed double-lift open-shed Jacquard machines with an additional needle device produced by Verdol and Zangs jointly, which enables an increased speed of up to 250 min^{-1} (rpm).

Figure 3.15 illustrates a Jacquard machine with the Verdol device (Fig. 3.15(a)) and the machine itself (Fig. 3.15(b)). The machine consists of 1344 needles (3) and 1344 double hooks (11, 12). Special features of the machine are:

- The double hooks (11, 12) have an extra projection (13).
- Two movable griffe blocks (knife frames) (14, 15), which perform, as usual, with reciprocating motions, as well as a further lower stationary knife frame (16).
- A rubber tension bar (ligament) is attached to each harness cord, instead of a lingo.

These features enable an open shedding movement as follows. If the needle (9) does not affect the needle of the weaving machine and leaves the hook (11) at rest, then one of its upper projections is gripped by the knife of the frame (14), which lifts the hook (11) together with the neck cord (17) and the harness cord (8), forming the upper part of the shed. At the same time, the projection (13) rises a little over the knife of trap board (16). If the interlacing pattern requires the thread dwell to be in the upper part of the shed, then the projection (13) is lowered slightly (5 mm) on the knife of the trap board (16). If it is necessary to drop the thread to the lower part of the shed, then the needle (3) should be pressed down by the needle (4). At the same time, the projection (19) will pass the knife of the trap board (16), and the hook (12) will fall as far as it can go to the stationary trap board (20).

If the thread dwell is required in the lower position, then the pressure of the needle (4) on the needle (3) is also necessary to deflect the upper projections of the retractors from the knives of the reciprocating working frames (griffes) (14, 15). The needles of the weaving machine have springs (21) supported by the needle board (22) to return the needles and retractors safely to the original position.

From the setting up of Jacquard machines to the production of new fabric patterns takes up to 8 weeks, which makes it difficult to meet the rapidly changing requirements of the market. Companies such as Bonas, Staubli and Grosse produce electronically programmable Jacquard machines, which enable manufacture of woven fabrics with repeats of up to 50 000 weft threads. Related data input devices also allow the pattern to be formed from several pieces. The reprogramming process has also increased several times over for Jacquard machines.

3.7 Comparative analysis of shedding devices

- Tappet mechanisms have the same design (up to 14 heald shafts) and a single working part – the tappet – performs both heald shaft lifting and lowering functions. This device can operate at a relatively high speed, up to 400–600 min^{-1} and is therefore suitable for manufacturing single woven fabrics (up to 6 warp threads with various interlacings) in large quantities.

- Dobbies have a more complex design (for up to 24 heald shafts), with the formation of the shed and the shed changing process each controlled by a specific part; therefore, dobbies operate at somewhat slower speeds, up to 300 min^{-1}, so these machines are appropriate for manufacturing fabric with more complicated structures (up to 24 warp threads with various interlacings).
- Jacquard machines are of more complex construction to enable individual warp yarn control for the formation of the shed and the shed changing process ; hence, these machines operate at relatively low speeds (up to 200 min^{-1}) for producing more complex woven fabric and carpets.

3.8 Questions for self-assessment

1. What is the purpose of warp shed formation?
2. What are the main parameters of a shed?
3. How can the neutral line (NL) of the warp shed be designated and established?
4. What is meant by a full, incomplete upper (top) or lower (bottom) shed?
5. Under what conditions is an uneven (irregular) shed formed?
6. Which type of shed is better: uneven or clear? Explain why.
7. Explain what is meant by levelling of the warp threads, and its phases.
8. How are tensions in the top and bottom branches of the shed caused? Why can it change during weaving involving repeat interlacings?
9. What is (i) a 'balanced' shed and (ii) an 'unbalanced' shed?
10. How can an unbalanced shed on a weaving machine be established?
11. How can the stretch deformation (or strain) of warp threads at shedding be calculated?
12. Assuming that a weaving machine can have up to 20 heald shafts, explain, with reference to different shed configurations, how the tension of a warp thread in the forward part of a shed depends on drawing it in the first, the tenth or the twentieth heald shaft?
13. Under what conditions is the equality of deformation of stretching and the thread tension in the forward and back parts of a shed achieved?
14. Draw diagrams of shedding cycles for warp threads with reference to any kind of combined interlacing.
15. How it is possible to reduce the magnitude and the amount of tension variation of warp threads in a cycle of shedding?
16. What are the advantages and disadvantages of the tappet type of shedding device?
17. What different kinds of tappets are used on weaving machines for driving the heald shafts? Give their advantages and disadvantages.

18. How does a dobby differ from a tappet type shedding device?
19. Classify Jacquard machines according to the density of the arrangement of holes on punched cards.
20. What is the capacity (or power) of a Jacquard machine?
21. What are the features of a tappet in a Hodson heald shedding device?
22. Describe the disadvantages of the tappet type shedding device on an Elitex (Kovo) pneumatic weaving machine.
23. What are the advantages offered by the Sulzer tappet type shedding device?
24. How is it possible to regulate (i) the position of a heald shaft relative to the height of the weaving machine, and (ii) the extent of its opening?
25. How does the profile of a tappet type shedding device relate to the cycle of shedding movements of a warp thread?
26. How is the sequence of a set of tappets dependent on the repeat of interlacings of the warp threads?
27. What are the main features of the Knowles dobby? Why is the heald shaft moved sharply on this dobby?
28. How does the Staubli dobby differ from the Hodson dobby?
29. How can the amount of movement of a heald shaft on the Staubli knife dobby be regulated?
30. What are the features of the Staubli rotary dobby?
31. What is the purpose of the development of the mechanical Jacquard?
32. Describe the basic design of a mechanical Jacquard.
33. Why does the basic mechanical Jacquard have a low operating speed, and what are the limiting factors?
34. What additional advantages are conferred by the Verdol device on a card type Jacquard machine?
35. How is the Verdol mechanism combined with the Zangs Jacquard?

3.9   References

1. Gordeev V.A. and Volkov P.V., 'Weaving', Leg. and Pitsh. Prom., Moscow, 1984 (in Russian).
2. Choogin V.V., Kahramanova L.F. and Nedovisiy M.N., 'Technology of Weaving Manufacture', State Technical University, Kherson, 2008 (in Russian).
3. Choogin V.V. and Chepelyuk E.V., 'The Forecasting of Manufacturability of Woven Fabric Structure', State Technical University, Kherson, 2003 (in Russian).

4
The supply of weft on the weaving machine

DOI: 10.1533/9780857097859.64

Abstract: This chapter considers how the supply of weft is maintained on a weaving machine. It describes methods and devices used for maintaining the weft supply on shuttle weaving machines, multishuttle mechanisms and shuttleless weaving machines.

Key words: weft detection, pirn changing, shuttle changing, weft change, shuttle and shuttleless weaving machines.

4.1 Introduction: the supply of weft on the weaving machine

The methods of supplying weft vary, as their names suggest, between shuttle and shuttleless types of weaving machines. The shuttle is the means by which weft is inserted on shuttle weaving machines. The shuttle carries a pirn (cop) on which sufficient weft yarn has been wound provide for the weaving of a few hundred picks. Each time the pirn runs out of weft, the shuttle has to be replaced with a fresh shuttle carrying a full pirn.

A manually controlled shuttle weaving machine has to be stopped in order to replace the shuttle. Shuttleless weaving machines provide higher weaving speeds, as the weft insertion on these machines is carried out using alternate weft insertion devices, which enables faster operation.

On older shuttle weaving machines, the replacement of the shuttle is carried out manually each time the weft runs out. This is inefficient since, usually, the machine has to be stopped each time a shuttle needs to be replaced. Such weaving machines are used in small-scale enterprises (such as family workshops, small handicraft establishments, etc.) and older factories established for the development of specific types of fabrics. To reduce machine idle time, the weaver usually replaces the shuttle when it is close to running out of weft. Given the relatively low speed of wide weaving machines, this operation can be carried out by an experienced weaver without stopping the weaving operation. At larger weaving mills, multishuttle weaving machines capable of the automatic exchange of pirns or shuttles are widely used. The automatic replacement of pirns and shuttles increases the productivity of weaving machines by between 25 and 200 per cent. Mechanisms for the automatic replacement of pirns and shuttles are used for change of weft of the same kind or colour (on single-shuttle weaving machines), or for the change of weft to a different kind or colour (as on multishuttle weaving machines). Automatic shuttle replacement mechanisms are less advanced than automatic pirn replacement mechanisms; hence, shuttle replacement mechanisms are used more frequently in the weaving of low linear density weft yarns (such as silk), as this requires less frequent shuttle replacement. Automatic pirn replacement involves four main devices: a gauge for detecting the depletion of weft in the shuttle, a pirn replacement mechanism, a battery (store) of pirns, and safety devices which guard against malfunctioning.

4.2 Gauges for detecting the presence of weft on shuttle weaving machines

The mechanical replacement of weft occurs in response to two circumstances: on a weft breaking, or on the depletion of weft in the shuttle. The weft feeler is a gauge that monitors the presence of weft on the pirn in the shuttle and signals its depletion. The detection of the presence of weft in the shed after picking is carried out by a weft fork (Ref. 1). These gauges free the weaver from having to supervise weft insertion closely and allow him to devote his attention to the operation of the weaving machine. Depending on the type of fabric being woven, the weft change operation is initiated by a weft fork or weft feeler. If the fabric is mass-produced in nature and fabric faults such as 'missing weft' are permissible, or when a broken pick can be easily separated and removed from the cloth fell, the weft fork can be employed to stop the weaving machine. If the fabric has close interlacing and when 'missing weft' is not acceptable, a weft feeler is employed to activate the weft change mechanism. The main function of the weft fork is to check the integrity of weft threads after their insertion in the shed (see Section 8.2 in Chapter 8).

Weft feelers (Fig. 4.1) can be of various types: mechanical, electrical, electro-inductive or photoelectric; they can also operate continuously or periodically. To permit the use of weft feelers, the pirn should have a reserve of between two and three picks of weft in order to avoid the weaving machine running out of weft. This, however, contributes to weft wastage, as it is necessary to remove the remaining turns from a depleted pirn before it can be rewound with fresh weft. Mechanical weft feelers monitor the presence of yarn on the pirn by the following methods: a feeler sliding along a contact surface (Northrop), penetration of the weft layer by a pair of electrical probes (Crompton & Knowles), the separation of weft coils (L.V. Zevakin), and by gauging the diameter of weft remaining on the pirn (Ref. 1 and Ref. 2).

Mechanisms of flat weaving technology

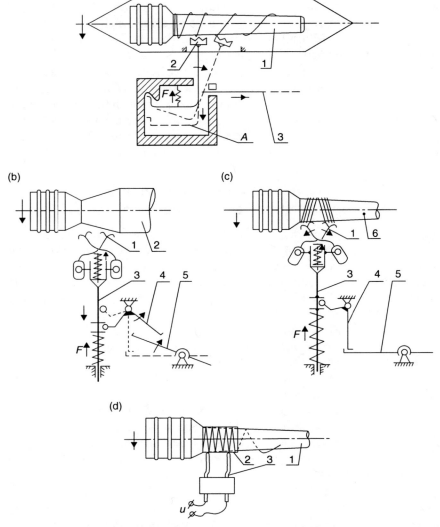

4.1 Weft feeler. (a) Slidings on pirn (Northrop); (b), (c) moving apart yarn coils (L.V. Zevakin); (d) electrical contact (Cohen). Refer to text for detailed explanation of components.

Sliding feelers are widely used, Fig. 4.1(a). The yarn on a pirn (1), moving with the shuttle, presses the head (pad) (2) of a feeler and moves it in the direction of position A. Signalling rod (3) remains at rest. At the depletion of the weft yarn, the feeler head (pad) (2) slips on the smooth surface of the pirn (1) and deviates to the right, moving a rod (3). This is the signal to activate the mechanism for pirn replacement. Thus, the feeler uses the difference of friction of the feeler head on the rough surface of the yarn and the smooth surface of the pirn.

The sophisticated mechanical feelers designed by L.V. Zevakin, shown in Fig. 4.1(b), separate coils on the depletion of yarn in the pirn. A yarn reserve is not required. A feeler head (1) on the yarn presses on pirn (2) and moves the core (3). A twin-arm lever (4) deviates from the line of action of a lever (5) on the pirn replacement mechanism. When the quantity of yarn on the pirn diminishes, the feeler heads (1) separate the coils on the smooth surface of a pirn (6). Displacement of the feeler heads (1) allows the core (3) and the lever (4) to remain static, which arrests the rocking lever (5). This activates the pirn replacement mechanism.

This way of detecting that a pirn needs to be replaced occasions a significant amount of weft waste owing to the difference between the diameter of the pirn and the amount of yarn needed to achieve the required minimum diameter. The electric feeler (Cohen, Fig. 4.1(d)) requires a special pirn (1) with a brass sleeve (2), or a special conductive varnish coating on the pirn stem. At a sufficiently depleted level of the weft yarn, contacts (3) are closed by the sleeve (2) (or by the varnish) on the stem of the pirn. Such feelers are used in silk weaving. The electro-inductive feeler requires a special pirn with a metal sleeve. The magnitude of inductance produced depends on the gap between the feeler and the sleeve, and is dictated by the thickness of a yarn layer. This type of feeler tends to produce considerable weft waste, especially with thin weft yarn. Of the different kinds of feelers used, possibly the most sophisticated is the photoelectric feeler. Its action is based on an estimation of the level of signal reflected from the surface of the yarn and the body of the pirn: the reflectivity of the pirn surface should contrast with the reflectivity of the yarn. A yarn reserve is not required.

4.3 Battery type weft supply

The quantity of the yarn on a pirn is between 40 and 50 times less than that on a weft package. On shuttle machines, six different types a pirn are used: round, tape, feeding hopper, vertical magazine, box and tape; these are used in combination with a moving Unifil head. The battery pirn supply (Fig. 4.2(a)), with a capacity of between 24 and 28 pirns, has been widely adoption (Ref. 1 and Ref. 2). Pirns (1) are loaded in a circular formation and release a fresh pirn by means of a transfer hammer (2), followed by the rotation of the battery to bring the next pirn into position. The pirns are held between a fixed disk and a keyed spring-loaded disk. The end of the weft from each pirn is supplied to the tooth of a directing disk and is reeled up on a stopper (3). This arrangement allows the end of the weft to be passed in the correct direction under constant tension. The battery can be configured

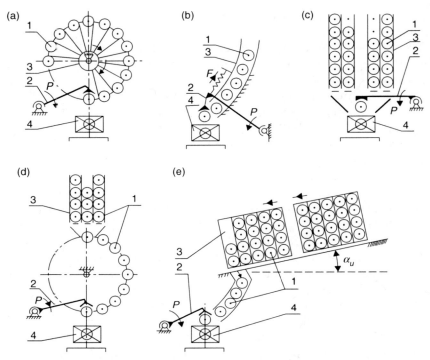

4.2 Battery of weft pirns. (a) Round; (b) ribbon; (c) feeding hopper; (d) vertical magazine; (e) box.

to carry pirns in two layers (Rüti). Each layer rotates independently, which permits the use of two different colours of yarn.

The tape weft battery (Fig. 4.2(b)) has a stationary section (3) and holds fewer pirns (1) than a circular battery. The supply system can also have two streams for pirns, permitting the use of two weft colours (Rüti). Basically, tape supplies are employed when tubular cops are used, as in the production of linen fabrics. Multishuttle machines use the Crompton & Knowles and Northrop type of feeding hopper (Fig. 4.2(c)). They have a larger capacity than a battery, but the process of drawing the end of weft threads in the shuttle during setting up is complicated. Usually, yarn of a different colour is put into each feeding hopper (3). To increase the capacity of the usual round or feeding hopper, a vertical magazine store is placed above it (Fig. 4.2(d)). This allows an operative to service two or three times as many machines. The ends of threads from the pirns are drawn and inserted manually. The Steinen box feeder, which makes use of weaving machine vibrations and the effect of gravity on the pirns, is set at an angle, as shown (G. Honegger, Rüti), which does away with the need for a special drive to displace the pirns along the slideway. In a modern system (Fig. 4.2(e)), the two boxes (3) contain 80 pirns each. This stock suffices for one working shift of the weaving machine. However, the wider use of these batteries is limited due to operational faults in the selection of the end of the thread from the pirns and the possibility of the ends of the pirns being damaged.

To increase the productivity of weaving machines, instead of using a battery of pirns, 'Unifil' winding (reeling) heads can be installed. For winding weft pirns, a winding head is provided with a tape battery for five full pirns. Automatic feeding of the winding head is carried out from weft bobbins. Cassette and box type batteries have sufficient capacity of pirns to supply a full working shift, but the construction of all known versions of these were not equally well designed, so occurrences of poor-quality pirn insertion were not unusual; hence, they tend to cause a high percentage of poor-quality pirn insertions in the shuttle. The best way to feed weft without requiring an operative is by using Unifil winding heads, which provide identical winding conditions of to all pirns, and hence high operational performance, leading to high- quality fabric.

4.4 Mechanisms for changing pirns

Replacement of an empty pirn should be carried out when the shuttle is in the shuttle box and remains motionless relative to the sley. Since the sley should itself be at rest to enable the weft to be changed, the optimum time is when the sley movement is in the dwelling phase. Sley dwell occurs at the extreme front and back positions of the sley. As the shuttle is moving through the shed when the sley in positioned as the back, a pirn can only be replaced when the sley is positioned at the front. Mechanisms for replacing pirns differ according to whether the transfer hammer is actuated by the movement of the sley (dependent movement), or is actuated by the main shaft (independent movement).

The Northrop mechanisms shown in Fig. 4.3(a), have become popular (Ref. 1). As the sley moves forward, the transfer hammer (2) is moved by picker (5), which is carried on the sley beam (race board) (3). The receiver (6) is hinged on the vertical shoulder (7) of transfer hammer (2). If there is weft on the pirn, the receiver spring is moved away by deflection from the line of action of the presser (5). In the event that it receives a signal to replace a pirn, the receiver (6), by means of the safety mechanism (8), is set on the line of action of the presser. With the sley moving into its forward position, the presser (5) strikes the receiver (6), moving the transfer hammer (2) clockwise relative to its axis of rotation. The head of the transfer hammer (2) moves the replacement pirn (1) to make contact with the empty pirn, located in the shuttle (4). The empty pirn is pushed out of the shuttle and falls into the open bottom of the sley and the replacement pirn is set in

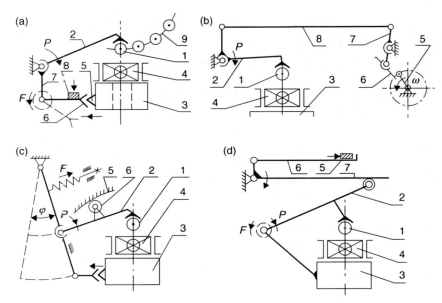

4.3 Pirn changing mechanisms. (a) Northrop; (b) B.I. Damaskin; (c) Rüti; (d) Zwicky.

place. The cycle of movement of the transfer hammer takes place over an angle of 46° of rotation of the weaving machine main shaft and is completed in the front dwell position of the sley.

In the B.I. Damaskin mechanism (Ref. 2), the cycle of movement during the replacement of a pirn by the transfer hammer (2) (Fig. 4.3(b)) is reduced to 24° of the rotation of the weaving machine main shaft. Transfer hammer (2) receives its drive from the main shaft (5) by means of a blow to a link (6) on the twin-shoulder lever (7). A percussive movement is carried out through connection (8) to the transfer hammer (2). The slow movement of the sley (3) during the replacement of a pirn increases the accuracy of pirn location. This operation should not be hurried as the resulting impulsive action of the transfer hammer may have an detrimental effect on the yarn on the pirn.

To increase the contact time of the transfer hammer with the pirn during insertion with the simultaneous coordination of the trajectory of the sley movement, V. N Krynkin (Ref. 1 and Ref. 2) has suggested the use of a profiled slope (5) (Fig. 4.3(c)) which acts on the roller (6) of the transfer hammer (2). The duration of pirn replacement is maximized (up to 48°) by the arrangement of the axis of the transfer hammer on the bar arm of the sley beam (Zwicky). In the mechanism devised by N.I. Asarov (Fig. 4.3(d)), a lever (5) moves to the right (see arrow), operating through the draught (6) on a two-shoulder lever (7), and lowers the transfer hammer (2), the axis of which is fixed on the sley beam (3). Despite its limitations, the Northrop replacement mechanism (Fig. 4.3(a)) has gained the widest application in shuttle weaving machines.

4.5 Safety devices of automatic pirn change

There are two kinds of safety device in pirn replacement mechanisms: devices to prevent insertion of the wrong pirn into a shuttle, and devices to avoid possible damage to components of the mechanism and the shuttle (Whittaker, Northrop, Diederichs); scissors (weft cutters) are also provided to remove weft ends that occur during pirn replacement to avoid flaws in the fabric that could arise by the weft ends being beaten-up (Ref. 1 and Ref. 2).

Because of its short duration (0.03 s), the process of pirn replacement is characterized as 'percussive'. If a shuttle is incorrectly positioned in the shuttle box when pirn replacement begins, damage can occur to the shuttle and to components of the automatic pirn replacement device. To avoid this, there is a safety mechanism that prevents incorrect location of the pirn. The safety mechanism (Fig. 4.3(a)) is connected to the receiver (6) by a ledge (8). When the signal from the weft feeler to replace the pirn is transferred to the sley, the accuracy of the shuttle position is established by means of the ledge (8) which positions receiver (6) on the line of action of the presser (5). If the shuttle (4) has not completely flown into the shuttle box sley (3), the safety mechanism, by virtue of the shuttle moving to its front position, lowers the receiver (6), and pirn replacement does not take place.

Rüti scissors, consisting of two sharp edges and one blunt edge. The weft of the old pirn is cut close to the fabric selvedge, and its free end is pulled out as the depleted pirn is ejected and falls into a box. The drive is derived from the safety mechanism by means of its interaction with rollers, stops, etc. Selvedge scissors cut off the ends of thread from the replacement pirn and the depleted pirn at the selvedge of the fabric. The drive to the scissors is provided from the sley by a cam on an intermediate drive shaft.

4.6 Change of hollow pirns

A hollow pirn (a pirn with a tubular structure) is used in weaving rough, thick linen or jute yarn. This type of pirn has no core and the thread is drawn from inside it, which imposes particular requirements for replacement of the weft. The ordinary automatic pirn replacement machine has the usual transfer hammer and tape battery. As the thread in the shuttle runs out, a signal initiated by the weft fork results in the automatic insertion of a pirn in the shuttle. Because the insertion of a full pirn is possible only once the previous cop is depleted, a 'no weft' weaving defect is inevitable. The firm Günne (Germany) produces automatic machines for replacing tubular pirns: the remainder of weft on a depleted pirn is ejected by a special lever before it is replaced with a full cop. This enables fabric to be produced without missing picks ('no weft').

4.7 Change of shuttles

Thin yarn, especially silk, is easily damaged by the action of the transfer hammer on insertion. Also, manufacturing fabric from natural silk thread (which is very non-uniform in thickness) demands mixing weft from different cops. Therefore, multishuttle mechanisms are also used on silk weaving machines. The difficulty of automating the replacement of massive shuttles with pirns holding yarn of various colours or different kinds of fibre is obvious. There are automatic weaving machines which can replace weft yarn by a change of shuttle with or without stopping the weaving machine. The design of the Tsudakoma (Japan) automatic weaving machine allows the replacement of shuttles with silk yarn without stopping the weaving machine (Fig. 4.4). This mechanism uses a lever mechanism to eject the shuttle. A Harriman type pusher is used (Ref. 2).

Shuttles (1) (Fig. 4.4) carrying the weft yarn move down into two bays in the battery (2). After the weft feeler clamp (3) signals the need to replace a shuttle, one stream of the battery and the shuttle descend by gravity onto a platform (4) which moves down on the action of a stopping trigger (5) and the system of levers (6, 7, 8, 9, 10) under the influence of the cam (11). A cam (12) operates through a roller on the lever (13) on which the drummer (14) presses the tooth (15) of the lever (16) which, in turn, by means of tension rod (17), lever (18) and link (19) moves the pusher (20) to the left. Thus, the shuttle (21(b)) carrying a full pirn, under the influence of the pusher (20), moves down the valve (22) of the shuttle boxes and lifts the back plate (24) of the shuttle boxes with chamfer (23). Note that a valve permits movement of an object in one direction only. By its further movement, the shuttle (21) pushes out the shuttle (25) on the sley beam (26) and is located in the shuttle box. The valve (23) falls, and the valve (22) rises, and the replacement shuttle is secured in the shuttle box. Under the influence of cams (11, 12) and spring F_2, the links return to their initial position.

4.8 Multishuttle mechanisms

Multishuttle mechanisms are necessary for the development of fabrics with weft of various colours, fibres of various structure or very non-uniform thickness (natural silk, yarn from a carded spinning system, etc.). A multishuttle mechanism of any design consists of one or two magazines with shuttle boxes for several shuttles. Shuttle magazines can be located on one or both

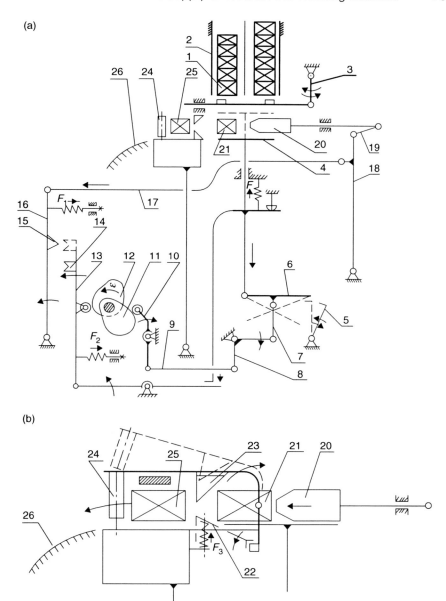

4.4 The mechanism for shuttle replacement on a Tsudakoma weaving machine. (a) The mechanism; (b) diagram of shuttle ejection, Hartmann.

sides of the sley beam and are moved relative to the sley beam by a special mechanism for replacing shuttles. The sequence of indexing of shuttles on the level of the sley beam is carried out by means of a card type programme carrier. Thus, depending on the arrangement of shuttle magazines, multishuttle mechanisms are divided into unilateral or bilateral (Ref. 2).

On multishuttle automatic weaving machines, the pirn replacement mechanism is co-ordinated with the operation of the multishuttle mechanism. On weaving machines with a unilateral multishuttle mechanism, pirn replacement takes place by means of an automatic mechanism, as on a single-shuttle weaving machine. On machines with a bilateral multishuttle mechanism, pirn replacement is manual. Depending on the arrangement of shuttles and characteristics of the movement of the shuttle magazine, shuttle replacement mechanisms have either a forward or rotary movement of boxes. In mechanisms with a shuttle magazine that moves forward, shuttles are positioned underneath one another and, for installation of shuttles on the level of the slay beam, the shuttle magazine makes backward and forward movements in a vertical plane. In mechanisms with a rotary movement of boxes, shuttles are positioned along the circumference of a drum that makes rotary movements when replacing the shuttle. These mechanisms are described 'revolving' systems.

Weaving machines with a forward shuttle magazine have been widely adopted. Depending on the characteristics of movement of a shuttle magazine, multishuttle mechanisms are divided into mechanisms with shuttle replacement denoted as random or consecutive. In random replacement mechanisms, the shuttle magazine can be moved by any number of boxes at

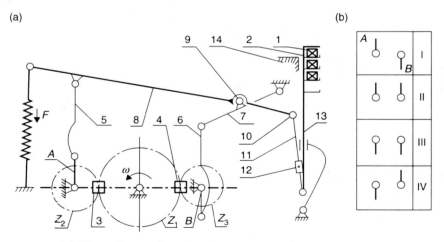

4.5 The unilateral four-shuttle mechanism (Leeming Hogdson). (a) The mechanism; (b) position of cranks *A* and *B*.

a time; and of the consecutive replacement mechanisms can only be moved by one shuttle box at a time.

Figure 4.5 shows a diagram of the mechanism of the Leeming Hogdson system, which operates the lifting of the shuttle magazine of the unilateral four-shuttle change mechanism (1) and is capable of the random replacement of shuttle boxes (2). The controlled lifting of the shuttle magazine (2) is carried out from a dobby by means of rods, levers and two movable teeth (3 and 4) which make randomly couple small wheels Z_2 and Z_3 with leading sector wheel Z_1. On the engagement of a movable tooth in the notch of small sector wheel Z_2, the large sector wheel Z_1 (fixed on the middle shaft of the weaving machine) turns the small wheel Z_2 at 180°, which turns the crank A fixed on it. Cranks A and B can occupy positions at the extreme top or the extreme bottom. The position of the cranks is changed by means of rods (5 and 6), a lever (7) is transferred to the elevating lever (8) and the axes (9) of lever (7).

The mechanism that operates the lifting of a shuttle magazine moves the hinge (10) of the elevating lever (8), connected with the shuttle magazine (2) by means of a rod (11) from the safety devices (12) and the bar (13),

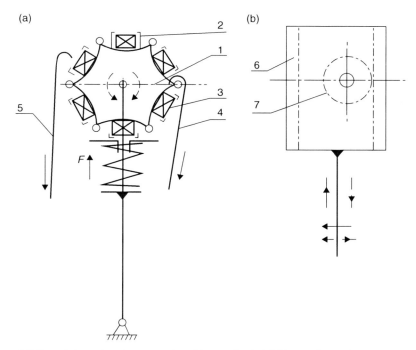

4.6 Revolving shuttle magazine (Hattersley). (a) The unilateral mechanism with consecutive replacement of boxes; (b) rotary device with the arbitrary change of boxes.

and has four possible positions relative to the sley beam (14). As a result, depending on the position of cranks A and B (Fig. 4.5(b)), one of four boxes of the shuttle magazine is positioned on the level of the sley beam (14). The Textima bilateral multishuttle mechanism (eight boxes for a maximum of seven shuttles) uses the same principle of operation as the Knowles dobby.

Figure 4.6(a) presents a Hattersley revolving shuttle magazine (1) with consecutive change boxes (2). The positioning of shuttles (3) on the level of the sley beam is carried out by two hooks (4 and 5), which are driven from cams. The operation of the hooks (4 and 5) is controlled by means of metal cards. A revolving box (1) can be turned by one box in either direction. To turn the shuttle magazine from between one to three boxes at a time, the bilateral rack (6) is alternately linked to the wheel (7) on the axis of the shuttle magazine (Fig. 4.6(b)).

4.9 Weft supply on shuttleless weaving machines

Distinctive features on shuttleless weaving machines are:

- the provision of several stationary weft yarn packages of a different colour or type as required by the fabric being woven;
- the requirement for presentation of the weft thread chosen from these packages to the weft picking device;
- the methods by which weft is unwound from the chosen package (Ref. 2).

According to the methods by which the weft thread can be presented to the picking device before insertion into the shed, the mechanisms for weft change (weft selection) can be divided into two groups:

- with the transfer of the weft to the picking device while it is at rest;
- with the transfer of the weft by the picking device at the beginning of its movement into the shed.

According to the method of weft thread wind-off during weft insertion, it necessary to identify two variants of mechanism:

- with periodic thread unwinding, from the 'rough' surface of a stationary weft package;
- with continuous winding from the surface of the weft package in combination with periodic weft winding from the smooth surface of the rotating drum of an accumulator type measuring device.

Mechanisms for weft change involve:

- A device for passing the chosen weft thread to the picking device (weft presenter); a battery of weft packages, each with an incorporated length measuring device (weft feeders) from which to select the required weft yarn.
- A battery of yarn consisting of package holders for mounting the required number of weft packages, sufficient for one working shift of the weaving machine.

For producing a fabric using weft yarns of different colours or of different fibres, additional mechanisms are used for the change of weft for transfer to a picking device before its insertion into the shed.

On the Sulzer projectile weaving machine, devices for changing weft of two or four colours or types (Fig. 4.7(a) and (b)) are used. The drive of the device is obtained from a disc (1) with fingers (2), resulting in the periodic rotation of a prism (6) with a metal control chain (7) by means of the Maltese cross (3) and wheels (4 and 5). The metal plates (7) have profiles such that the position of a roller (8) on the two-shoulder lever with a sector rack (9) is changed, which locate feeders (10) of weft (11) relevant to the picking device (projectile) (12(b)) before breakage. When chosen, one of the feeders (10), together with the end of the thread (11) clamped in it, moves to the picking device (12).

On rapier waving machines, the selection of the change of weft is carried out differently: the tensioned tip of the chosen weft thread is transferred from the relevant package to the head of a rapier on its way into the shed. Figure 4.7(c) presents one of variants of the weft selection mechanism (Iwer). Before reaching the head of the rapier (1), the end of the thread (2) coming from the lowered needle (3) is tensioned by air pressure from the nozzle (4). The rapier (1) grasps the end of the presented weft thread at the entry to the shed (5). The number of needles (3) corresponds to the number of weft packages. The vertical movement of needles (3) is selected by the programme device (6).

On air-jet weaving machines, the thread from each weft package (1) (Fig. 4.7(d)) is drawn and uniformly wound onto the smooth surface of the drum of a metering device (2) (also known as a 'weft accumulator' or 'pre-winder') and then fed to nozzles (3 and 4). Air is supplied to one of the nozzles from channels (5 and 6) in accordance with the programme device. The figure shows the use of two weft yarns, but up to eight or so can be used.

78 Mechanisms of flat weaving technology

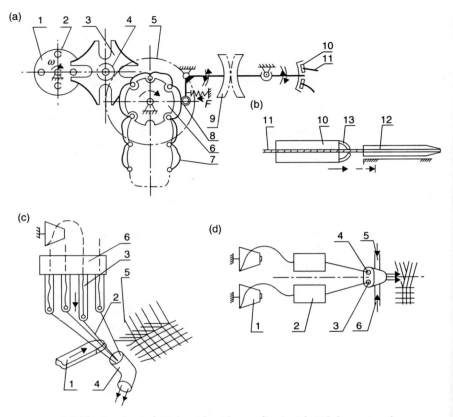

4.7 Mechanisms for changing the weft on a shuttleless weaving machine. (a), (b) Projectile (Sulzer); (c) rapier weaving machine; (d) pneumatic and hydraulic weaving machine.

4.10 Devices for measuring and control of weft tension

The high speed at which shuttleless weaving machine operate creates a need for high speed picking devices. If the weft thread is unwound intermittently and directly from the relatively uneven surface of a yarn package, sharp, dynamic tensions will occur due to sudden weft movement at the beginning of weft insertion. The at the beginning of its movement into the shed weft may reach a speed of 40 m/s on a typical air-jet weaving machine. The possibility of yarn breakage related to direct unwinding from the yarn package will be considerable.

Various methods are used to overcome the problems associated with unwinding a thread directly from a package (Ref. 2). For example, before the insertion of weft, a formation of a loop (1) is created (Fig. 4.8(a)) by a special

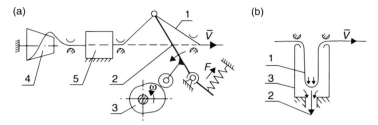

4.8 Weft tension compensators. (a) Mechanical; (b) pneumatic.

compensator (2) set in motion by a cam (3). The movement of the compensator (2), designated by the profile of the cam (3), corresponds to the manner of weft insertion of the type of weaving machine concerned.

A loop (1) (Fig. 4.8(b)) of weft thread can be formed by means of air pressure (2) created in a tube (3). Loop formation can also be achieved by frictional means using the moving rough surface of a closed belt. However, the length of the thread in the loop formed by the pneumatic and the frictional methods will vary. During the initial phase of picking the weft into the shed, the loop is used to decrease the dynamic tension in the weft caused by the picking device. In the second phase of its operation, the lever of the compensator (2) (Fig. 4.8(a)) occupies the bottom position (on the line of movement of the thread) and allows the thread to be drawn off the yarn package (4) through the tension device (5) in a straight line. Thus, compensators do not completely solve the problem of providing a constant tension in the picking phase of a weft thread, as the low resistance of a loop is combined with a subsequent jerk at the beginning of take-up from the bobbin.

A more rational variant of continuous wind-off of thread from the bobbin is by means of winding onto an intermediate measuring drum with a smooth surface, from which weft can be drawn off at the right moment without harmful resistance being caused by the pulling effort of a weft picking device. A considerable number of designs of measuring mechanisms combining the processes of measuring and accumulating weft during continuous taking-up from the bobbin have been invented. Figure 4.9 presents the basic variants of measuring mechanisms.

On ATPR (Russian) pneumatic-rapier weaving machines, the measuring device is used; this consists of a conical drum (1) (Fig. 4.9(a)), a clamping roller (2) and the store of a loop (3) containing a finger (4), supported on the lateral surface of a pulley (5). The cone of the drum (1) allows the regulation of the length of the weft in the shed, and the store (4 and 5) provides for the harmonic law of supply of a thread to the rapier (6).

Figure 4.9(b) shows the Maxbo-Murata device which measures a certain length of thread (1) which is wound up by means of a pressure roller (2) and

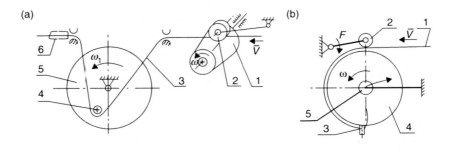

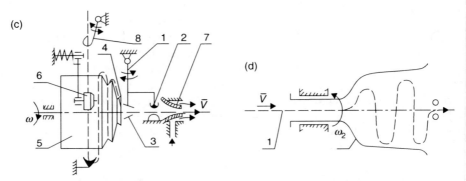

4.9 Weft measuring mechanisms. (a) ATPR; (b) Maxbo-Murata; (c) Kovo; (d) Elitex.

a needle (3) onto a smooth drum (4). At the beginning of weft picking, the needle (3) is withdrawn into the drum (4) and, under the influence of a current of pressurized air, the thread is easily taken up through the ring (5). At the end of weft insertion, the needle extends out of the drum (4) and stops the thread movement. The next cycle of winding (measuring and accumulation) of thread, with a set length for the full width of the reed, then repeats at successive weft insertions.

A reliable design of weft metering device, yet one that has a basic deficiency, is presented in Fig. 4.9(c) (Kovo). After the picking lever (1) goes to the left, and jams the weft thread into a clip (2), a funnel (3) simultaneously takes the thread to a tooth (4) of a spiral flute on a lateral surface of the drum (5). This initiates the process of winding the thread onto the drum by means of a clamping roller (6). A stream of pressurized air is supplied by the nozzle (7), following which a lever (1) moves to the right and releases the thread from the clip (2), subsequently removing the ring (3). Take-up of the thread begins with the drum (5). Then, a carrier (8) leads the thread out from under the roller (6) and take-up of the thread from the bobbin begins. Thus, the thread in the shed undergoes a dynamic jerk that often leads to formation of a 'bow' defect (the bending of the tip of the weft thread in the shed and a resultant flaw in the fabric).

An interesting design is presented in Fig. 4.9(d) (Elitex). Here, the thread (1) arrives at rotating chamber (2), is deposited by centrifugal force onto the internal wall of the chamber (2) in the form of a spiral, and the weft is taken up from inside the chamber by a picking device. However, coils of the thread can become tangled if the system stops.

Innovative methods are used for shifting yarn coils on the surface of the drum of the measuring device:

- In the drum, there is a drive to a number of cogged, narrow, longitudinal tapes on the surface of the drum; the thread, being reeled on the drum and the tapes, is moved by the tapes, which are collectively moved to the edge of the drum, releasing a space in which to form new coils (Sulzer).
- On the surface of the drum, there are a number of uniformly spaced longitudinal slots. Each slot carries a plate and, collectively, the plates are given a continuous but controlled wiggling movement which continually moves the yarn coils along the drum toward its nose so as to make space for new turns to be added on the drum (Vamatex). On some variants of measuring devices, photoelectric gauges are used to monitor the presence of coils on a drum surface.

The most advanced of these devices have measuring mechanisms meeting the following requirements:

- The exact metering of the required length of weft threads (±5 mm).
- Accurate distribution of separated coils on the smooth surface of a drum.
- Steady movement of coils on the surface of the winding drum without loss of separation of the adjacent coil distribution.
- Combination of continuous weft thread unwinding from the weft package, with intermittent winding off the drum.
- Independence of winding off the thread from the rough surface of the weft package and easy wind-off of accurately located coils on the drum surface by the pulling action of a weft picking device.
- Maintaining a low level of resistance to thread movement at the beginning of winding off the drum until weft insertion is complete.

4.11 Questions for self-assessment

1. What are the differences between the automatic replacement of weft on a weaving machine and the manual procedure for weft replacement?

2. What devices does the automatic weft replacement mechanism contain?
3. What are the different types of weft feelers available, their configuration and features? Give their advantages and disadvantages.
4. What are the features of the different kinds of weft storage feeders? Also, give their advantages and disadvantages.
5. What are the basic differences between the various weft replacement mechanisms?
6. What are the functions of the safety devices in weft supply mechanisms (cop replacing)?
7. Why is the replacement of a tubular cop problematic?
8. How is the replacement of shuttles carried out on the Tsudakoma weaving machine?
9. What are the main features of the two basic variants of movement of shuttles in multishuttle mechanisms?
10. What are the distinctive features of weft replacement mechanisms on shuttleless weaving machines?
11. Why does the need arise to provide a pre-measured length of weft threads on shuttleless weaving machines?
12. What devices used for pre-measuring and storing weft threads prior to weft insertion on shuttleless weaving machine are considered advanced?

4.12 References

1. Gordeev V.A. and Volkov P.V., 'Weaving', 'Leg. and Pitsh. Prom.', Moscow, 1984 (in Russian).
2. Choogin V.V., Kahramanova L.F. and Nedovisiy M.N., 'Technology of Weaving Manufacture', State Technical University, Kherson, 2008 (in Russian).

5
Weft insertion

DOI: 10.1533/9780857097859.83

Abstract: This chapter describes the different methods of weft insertion used on weaving machines. These include shuttle, rapier, projectile, air-jet and water-jet methods. Weft insertion based on electromagnetic, pneumatic-rapier, microshuttle, and weft inertia methods are also discussed.

Key words: weft insertion, shed, shuttle, microshuttle, rapiers, projectile, air-jet, water-jet.

5.1 Introduction: methods of weft insertion

Weft insertion is the action of inserting and conveying the weft through the open shed of a weaving machine. There are two basic methods of weft insertion:

1. *Periodic:* when picking occupies a specific phase of each rotation of the main shaft of the weaving machine (e.g. 90°–220°), during which the warp is opened to form a uniform shed across its full width, followed by the insertion of a single weft across it. The reed then moves forward to beat up the newly inserted weft. This action is repeated cyclically.
2. *Continuous:* when the process of picking occurs over the full rotation of the main shaft. Weft consolidation is carried out without the need of a reed.

Each of the methods above can employ shuttle and shuttleless methods of weft insertion along a straight line or a circular arc.

The shuttle used to be the main device employed by weaving machines which produce fabric by the periodic method of weft insertion. The shuttle carries within it a pirn which holds a stock of weft sufficient for weaving a few hundred picks (see Chapter 4). The shuttle moves back and forth through the warp shed, inserting one pick in each traverse, propelled by a picking mechanism located on each side of the warp. Each movement of the shuttle is as shown in Fig. 5.1 (Ref.1 and Ref. 2).

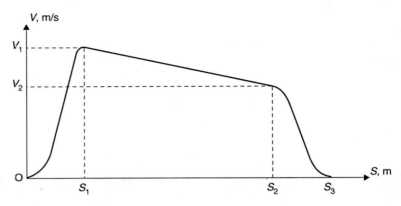

5.1 Shuttle motion on the weaving machine.

The interval $O - S_1$ corresponds to the period of shuttle movement within the shuttle box, during which its speed increases rapidly from 0 to V_{max} (V_1 = 13, ... ,15 m/s). The interval $S_1 - S_2$ represents the free flight of the shuttle through the shed. During this, a part of the shuttle's kinetic energy (ΔW) is dissipated as a result of overcoming friction against the reed and the lower warp threads of the open shed. The energy lost by the shuttle over part $S_1 - S_2$ can be represented as

$$\Delta W = 0.5 \cdot m_s (V_1^2 - V_2^2) \qquad [5.1]$$

where m_s is the weight of the shuttle with pirn.

For example, for V_1 = 13.5 m/s; V_2 = 11.8 m/s; m_s = 0.46 kg; we obtain: $\Delta W = 0.5 \cdot 0.46 \cdot (13.5^2 - 11.8^2) = 9.89 J$.

The amount of kinetic energy which the shuttle receives from the picking device is 5–10 times that which is dissipated by the shuttle flying through the shed. However, this magnitude of energy is necessary, since the time available for the movement of the shuttle through the shed is limited by the need to allow time for the shedding device to open the shed for the shuttle to pass through it. The shuttle flight should also coincide with the deceleration phase of the backward sley movement, when inertial forces keep the shuttle in contact with the reed and, hence, guide it correctly. During the third part ($S_2 - S_3$) of Fig. 5.1, braking of the shuttle occurs, bringing it to a halt from speed V_2 to 0, removing all of its remaining kinetic energy. The entire shuttle movement occurs in a very short time. A shuttle may weigh from 0.5 kg to a few kg, and accelerating it to speeds of the order of 10–15 m/s in a fraction of a second requires the the picking mechanism to exert large dynamic forces on it.

Where the pirn change is automated, the shuttle needs to be correctly positioned in the shuttle box under the set of pirns at the end of each weft insertion, taking account of changes to the pirn's weight resulting from yarn depletion and changes to its resistance on its flight through the shed. The exact setting of the shuttle is maintained by varying the braking force, according to the level of the shuttle's kinetic energy, using a stop rod which is pushed by the shuttle as it enters the shuttle box (Ref. 2).

5.1.1 Picking devices of shuttle weaving machines

Picking devices are subdivided into mechanisms with bottom (P_{tp}, Fig. 5.2(a)), middle (P_{mp}), or top (P_{up}) picking, according to where the main driving force is applied to the picking lever (Ref. 1 and Ref. 2). Consecutive or optional picking can be used, depending on the periodicity of the operation of the picking mechanism. The middle picking type devices have been most widely adopted on shuttle weaving machines (Fig. 5.2(c)). Picking devices are driven by a common shaft (1). On this shaft are fixed two picking cams (2), one at each end of the weaving machine, arranged at 180° relative to each other. On half a revolution of the common shaft (which corresponds to one full rotation of the weaving machine main shaft) the picking device moves the shuttle from left to right and, at the next half-turn, the second picking device moves the shuttle in the opposite direction.

Tappet, spring, crank or pneumatic type actuators are used to provide movement to the picking lever.

The picking cam (2) moves a picking bowl (3), carried on the spindle (4) of the picking shaft (5). On the same shaft, there is a connecting rod (6) which, by means of a small strap (7), a wooden bar (8) and a long strap (9), is connected to the picking stick (10). The picker (11) is attached to the top of the picking stick, and the tip of the shuttle (12) lies in the hollow of the picker. The side lever bracket (boot) (13) is attached to the bottom of the picking stick, resting on arm (14) on shaft (15). The arm (14) is set strictly parallel to a metal race board (16) of the sley. The side lever bracket (boot) (13) describes an arc of a circle with its centre in the hollow of the picker, converting the rocking movement to a parallel movement of picker and shuttle. This arrangement is necessary to propel the shuttle in the correct direction on its free flight through the shed.

In order to prevent the side lever bracket (boot) (13) from sliding along the arm (14), a hook (17) is fastened to the picking stick (10) which is carried on a roller (18). To return the picking stick to the starting position, a belt (19) is connected to the picking stick (10) and a block with a spring (20). At the instant of picking, the block is untwisted, and spring F twists up. After picking, the spring F reels up the belt (19) on the block (20), which returns the picking stick (10) to its initial starting position.

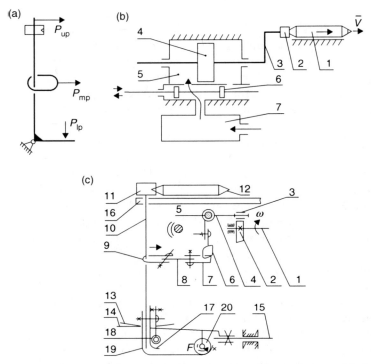

5.2 Picking devices of shuttle weaving machines. (a) Points of the application of driving effort to picking stick; (b) the pneumatic picking device; (c) typical picking mechanism. Refer to text for detailed explanation of components.

Each picking cam (2) maintains a constant radius over most of its rotation, which only increases sharply at its tip. This part of the cam is called the 'nose'. The force of picking depends on the height of the nose and the linkages of the picking device. The force of picking should be optimum: if it is too low, the shuttle cannot reach the opposite shuttle box; and if too high, its energy cannot be dissipated completely. Both extremes lead to incorrect positioning of the shuttle in the shuttle box, and to unsatisfactory operation of the mechanism, causing the weaving machine to stop.

The basic defect of this type of picking stick tappet drive is that the shuttle speed is dependent on the speed of rotation of the cam. This type of mechanism is therefore only applicable to narrow weaving machines. On wide weaving machines with a long dense sley and a heavy shuttle (0.5–1.5 kg), it is convenient to use a spring type picking device which relies on the picking stick for the stretched spring. The speed of the preliminary stretching deformation of the spring is not relevant, since the correct amount of stretching is the only requirement before picking can take place. In crank type picking devices, instead of cams or springs, a system of levers driven from the crank on the main shaft gives a sharp impetus to the picking stick, determining the length of weft picked.

Figure 5.2(b) shows the pneumatic picking device of the Novostav and Nopas metallic weaving machines (Czechoslovakia). The 570 g shuttle (1) is moved by the picker (2) and picking stick (3), which is set in motion by the piston (4) which charges the cylinder (5) with compressed air through a distributor (6). Management of the distributor (6) is carried out by a tappet drive. Air from a compressor first arrives in the receiver (7) for the stabilization of the compressed air entering the cylinder (5).

5.1.2 Positive drive of the shuttle

Wide weaving machines achieve movement of the shuttle mainly by the action of picking sticks. However, in narrow weaving machines the fabric can be narrower than the length of the shuttle. Lyall (Ref. 2) proposed a method of moving the shuttle using toothed gearing instead of the traditional free flight, which has been widely used in ribbon weaving machines (Fig. 5.3). Narrow woven tapes (ribbons) can be produced by a simple variation of the usual shuttle weaving machine design, involving a gear driver (1) (rack) (Fig. 5.3), gear wheels (2) and section reed (3). The number of tapes woven depends on the width of the weaving machine. At the back-and-forth motion of the rack (1), gear wheels (2) move a shuttle (4) through the shed. The beat-up of weft (5) is carried by a reed (3) in the usual way. After withdrawal of the reed from the edge of the tape (6), the shuttle (4) returns through the shed parallel to the reed (3). The weft is supplied by the pirn (7) in the shuttle (4).

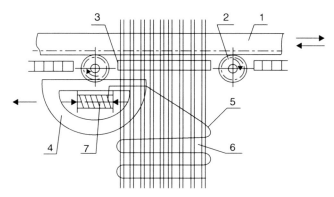

5.3 Shuttle drive of a ribbon weaving machine.

5.2 Continuous insertion of weft by shuttle

A method of continuous insertion of weft in a moving shed enables fabric to be woven by means of a 'point' beating-up system. Each length of weft inserted is continually pushed into the cloth fell, since no reed is used as in wide weaving. Continuous weaving of fabric occurs in circular weaving machines (Ref. 1 and Ref. 2), with the shuttle moving continuously at a constant speed on a closed trajectory, and fabric formed in tubular fashion. Several shuttles are moved simultaneously by mechanical or electromagnetic means for weft insertion, which is accompanied by a continuous process of shedding and insertion of weft into the cloth fell.

Figure 5.4 shows the Dynnika S.A.TKD-425L circular weaving machine which produces bags from rough linen yarn. The diameter of the fabric fell is 425 mm. Two shuttles are used for the weaving with a tubular pirn (wound from the inside). The speed of the machine is 200 picks per minute. Fabric is formed by the same operations as on the basic weaving machine, but these are performed consecutively rather than simultaneously. The warp yarns (1) (Fig. 5.4(a)) are divided into sections and come from two warp beams (2) with six sections of warp in each. The warp beams rotate synchronously by means of a special drive. On one of the beams, an automatic tape brake sets

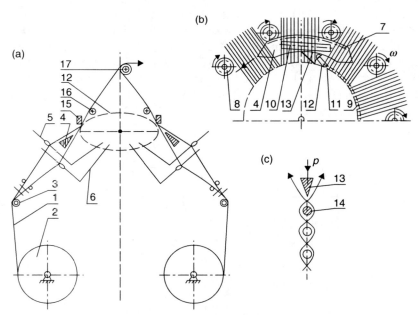

5.4 Circular weaving machine (S.A. Dynnik). (a) Weaving machine drawing; (b) shuttle movement; (c) consolidation of wefts in the cloth fell.

the necessary warp tension. The warp threads (1) bend around oscillating back-rests (3) which maintain warp tension, and pass through the apertures of the droppers and heald eyes of the harness framework (5) located in pairs in each section.

Two weft threads are continuously inserted by means of two shuttles (4). For one rotation of the main shaft of the carriage, the harness (5) moves one full cycle, carrying out shed formation continuously as required for each shuttle. The shuttles are moved in a circular path by gear teeth (7) (Fig 5.7(b)) on the shuttle, which engage with textolite gear wheels (8). While still engaged with one textolite gear wheel, the teeth on the shuttle engage with the following gear wheel, resulting in circular movement at a constant speed. Weft (9) is taken up from a tubular pirn (10), leaves the thread quill (11) and is laid along the cloth fell of the fabric (12). The condensing beater (13), which is in front of the thread quill (11), presses on the warp (Fig. 5.4(b)) now overlapping the the previously laid weft (14) and moves it into the cloth fell (12). The cloth fell (12) is aligned by the continuous ring breast beam (15 (a)) and spacing ring (16). The tubular fabric is flattened as it passes around the emery roller (17) and arrives at the take-up roller.

In the Sagem circular weaving machine (France), shuttles are set in motion by frictional contact with rubber rollers. However, with this method the movements of the shuttle and the harnesses may fail to coincide due to slippage between the rollers and the shuttle.

In the Fayolle-Anset weaving machine (France), eight shuttles are moved in the shed by an electromagnetic field created by a coil mounted at the end of a lever, moving along the fell of the fabric in the shed across the warp threads. In the Fokke weaving machines (Germany), the shuttle has a built-in electric motor. The shuttle rolls on wheels on the reed at the fabric fell.

A synthetic film split into strips and placed directly in front of the back-rest can be used as the warp. The main limitation of circular weaving machines, owing to the point beat-up of weft, is the density of yarn and, hence, the range of fabrics that can be produced. Basically, these weaving machines are used to produce packing or technical types of fabrics. Circular weaving machines achieve a rate of 300–1800 m of weft insertion per minute.

5.3   Weft insertion by microshuttle

The microshuttle is a device for inserting weft threads with a limited capacity of one pick, which makes it compact (Ref. 2). Microshuttles of various designs are used in multiphase flat weaving machines. They are guided in a shed by profiled levers bearing, in turn, on rollers or chamfers of the cases of

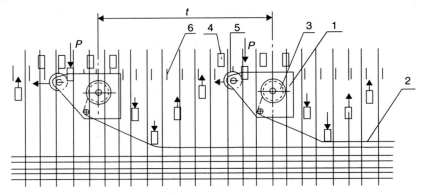

5.5 Microshuttle drive (Cerdans).

the microshuttle. A microshuttle can be moved by profiled levers, and along the cloth fell by rollers or by an electromagnetic field.

In Fig. 5.5, the microshuttle (1) movement in a wave shed in the Cerdans multiphase weaving machine (Spain) is shown. The term 'multiphase' here indicates that the picks at the fell are at different stages of the insertion process. The weft thread (2) is taken up from the disc spool (3) and pushed into the cloth fell with special plates (4) ('point beaters'). The plates bear on the microshuttle roller (5), moving it along the clothfell. The microshuttle slides along stationary plates (6), guided by a profiled slot on its body. The microshuttles are 250 mm apart, and they move in the shed at a speed of about 2 m/s. The weaving machine is bilateral, and each side is set to make 20 rectangular pieces of fabric, after which a special weft-reeling mechanism loads fresh microshuttle spools.

Figure 5.6 shows the elements of the experimental multiphase flat weaving machine by D.I. Popov (Ref. 2). Oscillating plates (2) press the edge of a circular microshuttle (1(c)) in a running wave shed, applying an impulse of movement along the fell of the fabric. The plates perform the movement of the weft packages, as well as the beat-up (consolidation) of the weft into the fabric fell. The microshuttle is low ($H = 6$ mm) and narrow ($D = 40$ mm) with no moving parts. When it leaves the shed, each microshuttle is reloaded with weft. The wave shed (Fig. 5.6(d)) is formed in a novel way with a moving chain (3) with rollers (4) and a flexible steel tape (5) to which harnesses (6) are attached. Microshuttles are spaced at t = 150 mm. The machine operates at 300 cycles per minute. Microshuttles in the MT-330, TMM-360 weaving machines (Russia) move in their respective sheds by the fabric forming mechanism at a speed of 1–2 m/s. The productivity of the weaving machine is 600 weft insertions per minute.

In the Rossman flat multiphase weaving machine (Fig. 5.7(a)), the weft is drawn from the bobbin (1), through a tube (2) which, by its rotation, reels

Weft insertion 91

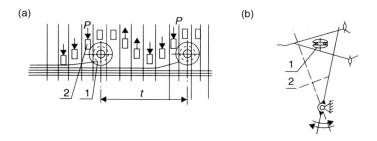

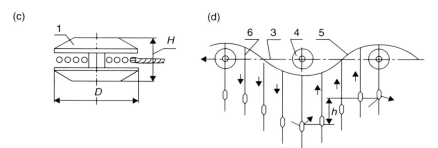

5.6 Elements of multiphase weaving machine (Popov).
(a) Arrangement of microshuttles in the shed; (b) beating plates;
(c) a microshuttle; (d) basic wave shedding design.

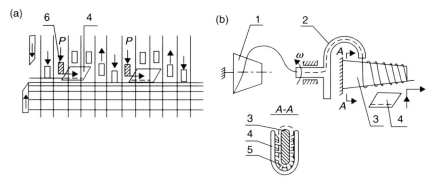

5.7 Picking of weft on multiphase weaving machine (Rossman).
(a) Winding the weft and drive of microshuttles in the shed;
(b) microshuttle.

the thread up onto a stationary wedge-shaped flat plate (3), which causes the compressed thread to slide along the plate towards the shed, overcoming the challenge of automatic transportation of coils of threads on the plate (3). The microshuttle (4) (Fig. 5.7(b)), in the shape of a bracket with bristles (5), approaches the wedge (3), captures the weft and moves into the shed. Oscillating plates (6) bear on the chamfer of the microshuttle (4) and move

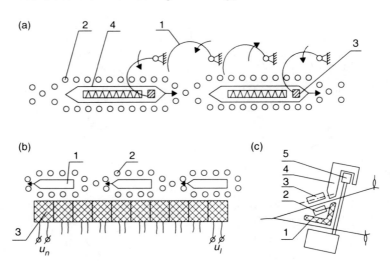

5.8 Variants of microshuttle drives. (a) Levers; (b) electromagnetic running field; (c) press rollers.

it along the cloth fell, pressing down a short element of weft into the cloth fell ('point' beating-up).

Microshuttles can be propelled by special bent levers (1) (Russia, Fig. 5.8(a)) the ends of which enter the shed between warp threads (2) and press the ledge (3) of the microshuttle (4). An electromagnetic running field for moving microshuttles (1) (Germany, Fig. 5.8(b)) can be created in a running shed (2) by consecutive switching of stationary coils (3), or by means of an electric chain moved along a sley bar under the shed. A microshuttle (1) (Fig. 5.8(c)) can also be moved by a driving roller (3) on a roller (2) on the microshuttle. Driving roller (3) is carried on a cap (4), moved along the top rib of the reed (5) by a chain. In the E.A. Onikov, TCP (Russia) and Pignone-Smit (Italy) multiphase machines, a loop chain with rollers moves at a speed of 2.1 m/s under the warp threads along the microshuttle rail. With a working width of 3.3 m, the loom has a weft insertion rate of 2200 m/min, producing fabric at 66 m^2/h.

5.4 Weft insertion by projectile

Numerous attempts to increase the speed of shuttle movement in a shed have led to the creation of a 'passive shuttle' (projectile) which does not carry a pirn, but is equipped with a clamp to capture the tip of the weft thread (M. J. Dantzer). Here the shuttle has been replaced by an inertial picking arrangement. The projectile (also known as an Inertial Small Picking Device – ISPD) is a device for inserting a weft thread into the shed without

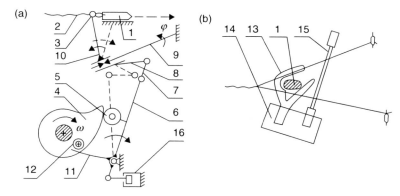

5.9 The picking device of a weaving machine (Sulzer). (a) The drive mechanism of the picking device of a weft; (b) The comb arrangement in the shed.

using a constant kinematic drive. Mechanical, pneumatic or electromagnetic devices can be used as picking mechanisms (Ref. 2).

The most widespread mechanical picking device of this kind is the torsion bar type drive of the Sulzer weaving machine (Ref. 1 and Ref. 2). Figure 5.9 shows the basic design of this picking device. The 40 g, 6 mm thick and 90 mm long projectile (1) has a clamp (clip) into which the weft tip (2) is transferred before picking starts. The projectile moves only from left to right, returning to its starting position through a special conveyor located under the shed. On this weaving machine, depending on its weaving width (1.8–3.2 m), between 9 and 17 projectiles move serially on a conveyor back to the picking device, enabling a high pick insertion rate.

The cam (4) (Fig. 5.9(a)), through the cam follower (5) on lever (6), is linked (7) to lever (8), twisting the torsion bar (9) by 27°–32°. The picking lever (10) moves to the left in the starting position before picking. As the shaft (9) finishes twisting, the profile shoulder (11), the lever (6) and link (7) are critically unbalanced. With the further rotation of cam (4), the roller (12) carried on cam (4) moves the shoulder (11), unbalancing the lever system and causing the torsion bar (9) to untwist, with a resulting sharp movement of the picking lever (10) towards a shed. The projectile (1), receiving kinetic energy from picker (3) in approximately 0.005 s, makes a free inertial flight through the shed, guided in a channel defined by profiled blades (13) (Fig. 5.9(b)), fixed on sley (14) in front of the reed (15). At the end of its flight, the projectile enters a receiving box, moves onto the conveyor and back to the picking unit to capture a fresh weft end and start a new cycle of picking. At the end of picking, the movement of the mechanism is dampened by a dashpot (16) (Fig. 5.9(a)).

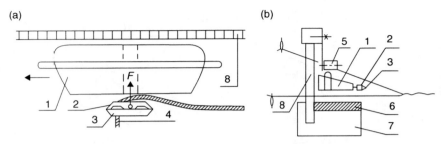

5.10 Weft insertion on the Textima weaving machine (Danzer). (a) Weft-picking device; (b) picking device in the shed.

In Neumann's Textima weaving machine system (Fig. 5.10(a) and (b)), the M.J. Dantzer 'passive shuttle' concept is used. In this weaving machine, the weft-picking device is a flat case (1) 12 mm high, 33 mm wide and 160 mm long, containing a symmetrical clip (2) with a sharp edge (3). The clip (2) presses against the case (1) with a system of springs F. The tip of the weft thread (4) is captured at the beginning of the case's movement before it enters the shed. On leaving the shed, the press roller (5) opens a clip (2) on the case (1) and releases the tip of the weft thread. The picking device, which has a metal picking stick with a rubber picker, flies in alternate directions to pick the weft. It is guided in the shed by the race board (6) of sley bars (7) and by a reed (8).

In the Novostav weaving machine, the picking device is moved by a pneumatic mechanism and is guided in the shed by directing needles and the reed. After leaving the shed, the picking device turns back through 180° to pick in the opposite direction. The reed becomes worn through contact with the picking device's steel case, which increases the rate of warp thread breakage. The same defect is likely to occur in the Neumann system Textima weaving machine.

To speed up a microshuttle (1) (Fig. 5.11), it is possible to use an electromagnetic field created by a stationary coil (2). When power is applied to the coil (2), the magnetic field in the coil (2) attracts a steel ISPD (1) which has the tip of the weft attached to it. The picking device is thus accelerated till the case reaches the required speed. When the current is switched off, the picking device (1) flies freely in the shed. The receiving coil (3) is located at the exit of the shed and the electromagnetic field applies a brake to the movement of the picking device (1). The picking device is returned to the same starting position by means of a special conveyor located under the shed. The contactless method of accelerating and retarding the picking device enables durable operation of the mechanism.

As an exception to use of the pulsed inertial ISPD, Cuper proposed the use of moving electromagnets to convey the ISPD in the shed under the constant influence of a magnetic field. In his circular weaving machine, 12 ISPDs continuously move, so that weft yarn of the required colour can be presented in the path of each ISPD for capture and insertion, making it possible to produce checked fabric.

5.5 Weft insertion by rapiers

In shuttleless weaving machines, the the rapier method of weft insertion has been the most widely adopted. The rapier is a weft insertion device which avoids the need for a moving weft package or pirn, by maintaining constant kinematic communication with a drive. Both intermittent and continuous weft insertion by this method are possible (Ref. 2). Intermittent insertion of weft threads by rapiers requires the following:

- rigid rapiers in the form of tubes, hollow cores, of constant or variable (telescopic) length;
- flexible rapiers in the form of perforated tapes, which are wound in and out of the shed on disks or segments;
- unilateral or bilateral rigid rapier mechanisms, depending on the arrangement of the shed on the weaving machine;
- thread picking as a loop (Gabler's method) or by the weft tip (Dewas' method);
- thread picking by a rapier to the middle of a shed and transfer to the opposite rapier;
- thread picking by a rapier over the full width of the warp;
- picking in the shed in one direction or both directions;
- installation of a rapier mechanism on the sley, or on the weaving machine frame.

Figure 5.12 shows the basic methods of rapier picking. In the Gabler method (Fig. 5.12(a)), the new pick (1) is first inserted by the rapier (2) in the form of

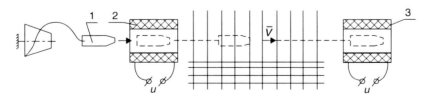

5.11 Electromagnetic picking device.

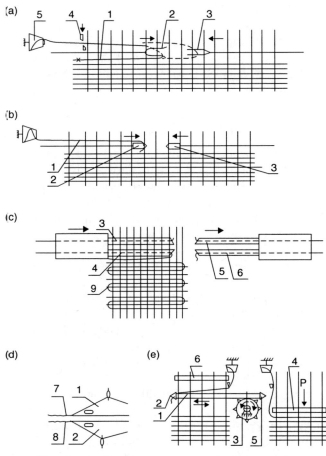

5.12 Methods of picking by rapier. (a) Gabler; (b) Devas; (c), (d) Tumack; (e) two-trailer rapier (Saurer).

a loop in the middle of the shed, where it is intercepted by a receiving rapier (3). The weft loop is opened by scissors (4) at the insertion side, and the weft is straightened by a hook on the rapier (3) as it moves to the far side of the fabric. The disadvantage of this method is that the thread is withdrawn from the weft package (5) at double speed and the weft may be cut on its transfer from rapier (3) into the tip of rapier (2).

Under the Dewas method (Fig. 5.12(b)), the thread (1) is moved to the middle of the shed by its tip by the rapier (2). The head of the receiving rapier (3) intercepts the end of the weft and takes it to the right-hand edge of the shed. There is a possibility that the weft may be cut by the contact with the rapiers.

According to Tumack (Fig. 5.12(c) and (d)), double rapiers (3, 4) and (5, 6) insert two picks simultaneously in two sheds (1, 2) to form two fabrics (7, 8). Weft in the fabric settles down in the form of 'hairpins' (9), and a selvedge-forming mechanism is not therefore required. Each double rapier inserts weft serially over the width from the two sides of the warp.

The company Saurer has offered, for economic reasons, a method of weaving fabric using a rapier (1) with two heads, or tips (2, 5) in the left and right shed, respectively (Fig. 5.12(e)). While the left head (2) of rapier (1) picks the weft thread in the left shed with the help of gear wheels (3), the right reed (4) beats the previously picked weft into the clothfell in the right shed. Then the rapier leaves the left shed and the right head (5) inserts weft from the other bobbin into the right shed. When the rapier exits, the reed (6) beats the weft into the fell of the left-hand length of fabric.

Figures 5.13(a)–(f) present several methods of rapier drive.

(a) Lever with a drive from a crank (1) (Iwer).
(b) Lever with a drive from a cam (1), controlling the hinge A of rigid rapier (2) which is given a rectilinear movement by means of the system of links (3, 4, 5 and 6). The rapier (2) is fastened on the continuation of the sley (7) on the sley shaft (8) (SACM).
(c) Rack-and-pinion (gear) in which gear rod (1) is set in motion by a crank on the main shaft and, in turn, by means of gear wheels (2, 3) turns a disk (4) which carries the end of the flexible rapier (5) bearing the head (6) of a weft thread (Draper).
(d) Tappet gear in which the rapier (1) moves back and forth by the use of a flexible punched tape (2), driven by a grooved cam (3) by means of gear sector (4) and gear wheels (5, 6) (Nuovo Pignone, Snoeck).
(e) Screw, in which the crank (1) moves the screw (2) back and forth, resulting in rotation of a nut (3) and sector (4) on which the tape (5) with the head of the rapier (6) is carried (Vamatex).
(f) Frictional, where the rapier (1) in the form of a flat steel tape with gripper (2) circulates in one direction at a constant speed by means of two continuous belts (3, 4), between which the tape (1) is clamped, with disks (5, 6) providing the drive. The rapier (1) in the shed (7) is guided by directing plates (8) (D.V. Titov).

Comparative analysis of available data shows the following:

- The weight of rapiers with heads does not exceed the weight of a shuttle (up to 300 g).
- Overall dimensions of a weaving machine with rigid rapiers can be reduced using telescopic rapier designs (Galileo).

98   Mechanisms of flat weaving technology

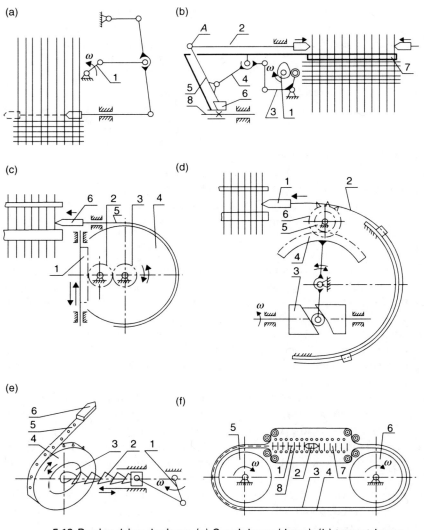

5.13 Rapier drive designs. (a) Crank-lever (Jwer); (b) tappet-lever (SACM); (c) jagget-rod (Draper); (d) tappet-gear (Snoeck); (e) screw (Vamatex); (f) frictional (T.V. Titov).

- Deflection of rigid rapiers in the shed by the reed and warp make the exact transfer of a thread by the rapier heads somewhat unreliable; however, the fact that directing plates are not needed makes it possible to develop warp-intensive fabrics.
- Rigid rapiers impose a limit on the the drawing-in width of the weaving machine.

Weft insertion

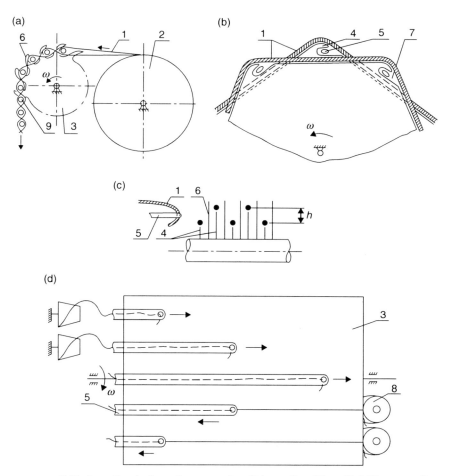

5.14 A method of continuous of weft insertion (Gentilini-Ripamonti). (a) Reading by a warp; (b, c) shedding disks; (d) movement of rapiers.

- Flexible rapiers allow fabric more than 5 m wide to be produced, but rapier guide combs are needed in the shed.
- Rapiers driven back and forth cause high dynamic loadings that restrict the speed of the weaving machine.

Multiple rapiers are used for continuous weft insertion in the Jentilini-Ripamonti weaving machine (Ref.1 and Ref. 2) (Fig. 5.14). Warp threads (1) (Fig. 5.14(a)) are drawn from the warp beam (2) on a circular drum (3). In the shed (7) (Fig. 5.14(b)), formed by displacing warp threads (1) by the edges of shedding disks (4), weft yarns are inserted by rapiers (5) from bobbins. Two kinds of disks rotate the drum (3): shedding disks (4)

(Fig. 5.14(c)) and beating disks (6). The latter also keep warp threads correctly spaced. In their lateral grooves (7) (Fig. 5.14(b)), more than 20 rapiers (5) move with a phase lag (Fig. 5.14(d)). On arrival at the edge of the drum (3), the end of the weft thread is clamped by a rubber roller (8), and following that the rapier (5) leaves the shed to take up its starting position. Beating disks 6 (Fig. 5.14(a) and (c)) press every weft into the cloth fell (9), which is taken up by a drawing pair of rollers and arrives at the taking-up roller. The thickness of the disks (4) corresponds to the thread thickness. The rapiers move at a speed of 7–8 m/s. The drum (3) rotates at 100 rpm. The productivity of the weaving machine is 1000–2000 weft insertions per minute.

5.6 Weft insertion by air and water jets

To remove restrictions on the high-speed operation of the weaving machine, researchers have developed methods of weft insertion which eliminate the extra time and energy expended in driving a weft-picking device (such as the rapier method) and the dynamic loading imposed by weft insertion mechanisms.

5.6.1 Weft insertion by air jets

The main problem encountered in bringing the weft into the shed by a current of air is the fast dispersion of the core of the airstream and the loss of traction on the thread (Ref. 1 and Ref. 2). To increase the reliability of weft insertion at widths above 45 cm (P-45, Kovo) plate-limiters (1, 2) (Maxbo) are used to conserve the airflow (Fig. 5.15(a)), which can increase the length of the weft pick to 100–125 cm (P-125, Kovo). The picking length is increased when a series of profiled plates (1) (Fig. 5.15(b)) which have an internal conical surface (2) (Fig. 5.15(c)) are mounted on the sley bar (4). Plates (1) (Fig. 5.15(b)) form a channel which conserves the air flow. While the plates (1) move under the cloth fell (5) as the sley moves forward during beat-up, the weft (6) moves out of the tunnel through a groove (7) and remains on the bottom warp sheet, to be beaten up by the reed.

The basic method of pick insertion by a fluid (water or air) is shown in Fig. 5.15(d). The thread from the bobbin (1), having passed the thread tensioner (2), is taken up from the metering device (3) and arrives in the nozzle (4). Compressed air moves in a diffuser (5) and, flowing in the outer nozzle (4), comes into the shed, grasps the end of the thread (6) and picks it into the channel defined by the plates (7). To achieve an increase in the length of the pick inserted, additional nozzles (8) are set at certain positions along the channel. At the exit from the shed there is a stretching nozzle (9), whose function is to suck in the tip of the weft and keep it

Weft insertion 101

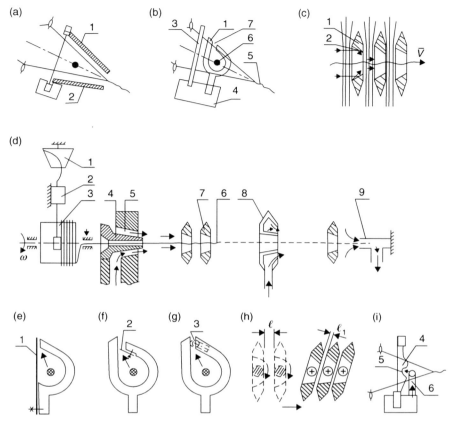

5.15 Inserting weft by air or water jets. (a) Maxbo-Murata; (b), (c) and (d) Kovo; (e) Talio Gialoni; (f) Jettis; (g) J.I. Syromjatnikov (PT-175); (h) rotary plate-limiters; (i) Te Strake.

straight during beat-up. Thus, the length of weft picked can be increased to between 2 and 3 m.

At the same time, plate-limiter design has been improved to reduce their thickness and the backlashing effect on the weft. The groove on the weft exit from the channel can be closed by a spring plate (1) (Fig. 5.15(e), Talio Gialoni), a springing uvula (2) (Fig. 5.15(f), Jettis), or a ball (3) (Fig. 5.15(g), PT-175, Syromjatnikov J.I.) The latter is in a recess at the edge of the plate-limiter at the exit of the thread from a shed. The channel can be consolidated by periodic tilting of the plate-limiters in the weft-picking phase (Fig. 5.15(h)) when the gap ℓ decreases to ℓ_1.

In the Te Strake weaving machine, instead of plate-limiters, a profiled reed (4) is used (Fig. 5.15(i)) with the channel ('tunnel') (5). Weak conservation of the current of air in the channel requires relay needle nozzles (6) to be installed at 50 mm intervals along the reed. Air consumption is essentially greater than on weaving machines with plate-limiters. In modern pneumatic weaving machines, the speed of air issuing from nozzles reaches 300 m/s, and the initial speed of the tip of the weft is around 35–50 m/s.

5.6.2 Weft insertion by a water jet

The basic water-jet method of weft insertion is identical to that shown in Fig. 5.15(d). Water under pressure arrives at an atomizer (5) and, flowing round the nozzle (4), enters the shed in the form of a stream, grasping the tip of the thread (6) and laying it in the channel. The water stream from the nozzle starts to divide at a distance of 15 cm, and at 50 cm it turns into a drop-shaped cloud. The wet fabric is dehydrated in the weaving machine by a squeezing shaft (H-145, Elitex) or by sucking up the moisture through a slit in the hollow breast beam (Nissan). However, not less than 30% of the moisture remains in the fabric, necessitating drying outside the weaving shop. The water consumption in the weaving machine is between 25 and 50 litre/hour, or 0.5–1.5 g per single pick, with a pressure of 0.05–0.15 Mpa and a temperature of 16–24°C. Water-jet weaving machines are basically used for the production of fabrics from synthetic yarns.

5.6.3 Features of weft insertion by air-jet and water-jet methods

The analysis of results from the operation of jet weaving machines and experimental research highlight the following features of these weaving processes (Ref. 2).

Advantages of the use of a fluid stream include:

- No dynamic loading of the weft insertion mechanism, since the only moving part is the control valve supplying the fluid to the nozzle.
- The high picking speed increases the working speed of the weaving machine to over 400–600 wefts per minute with weaving width from 1.25 m (Kovo) to 3.0 m (Tsudakoma).
- High operating ratio (utilization) of manufacturing space.

Disadvantages include:

- Higher rejection rate of the weft (3.5 per cent), owing to unstable weft insertion conditions.
- Occurrence of certain types of defect in the fabric, such as short weft and lacing, uneven fabric formation on the left and right edges (weft undergoes some untwisting during flight; in water-jet weaving the fabric is less damp at the nozzle).

Reasons for the formation of defects in the fabric include:

- At lower pressures of the stream, the weft does not fully leave the shed ('short weft').
- Excessive pressure can cause breakage of the weft.
- With an imperfect shed, sagging of at least one warp thread is likely, and this can cause stoppages of weft against the walls of plate-limiters or the profiled reed. This may prevent the tip of the weft from moving ('lacing' faults).
- Uneven weft twisting can cause low twist areas, as the weft end untwists to some extent during its flight.
- Fluctuation of air pressure in the centralized pressure head network.
- Contamination of flow by dust or oil.

The following recommendations may reduce fabric faults:

- Using a process of *relaxation* of strain in the threads to achieveeven twist *before* use on the weaving machine.
- Ensuring a high degree of cleanliness of the stream by filtration of dust and oil particles. Not allowing air pressure to fluctuate significantly in the centralized network, which may occur when the number of simultaneously working weaving machines changes unpredictably.
- Providing optimum air or water stream pressure.

5.7 Pneumatic-rapier weft insertion

To reduce the problems of pneumatic and bilateral weft insertion (contact of rapier heads, complexity and multipart nature of the devices, fast dispersion of the air current over the length of picking and loss of traction effort on the weft, high levels of waste of air, etc.), the combined pneumatic-rapier method (ATPR, Russia) was created (Ref. 1 and Ref. 2). The pneumatic-rapier method has the following two basic advantages:

1. The ends of the rapiers do not make contact in the shed when transferring weft.
2. There is no dispersion of the pressure head, as the flow of air in the rapier tube allows a reduction in air consumption.

Figure 5.16 shows the arrangement of rapiers and the compressed air supply. The basic method of operation is that the rapiers enter the open shed and their tips (ends) come close together to enable the weft to be inserted through the right-hand rapier. From the central compressor station, the compressed air

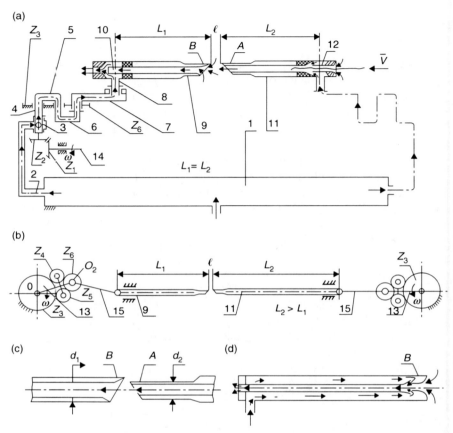

5.16 Pneumatic-rapier method of picking. (a) Air circuit; (b) rapier drive; (c) rapier heads; (d) suction by rapier.

arrives through the receiver (1), passes through channels 2–8 on the left to the target end (not to head *B*) of the receiving (taker) left-hand rapier (9). The air stream, flowing round the nozzle (10), exits the rapier to the left, creating a suction stream in the rapier (9). Compressed air arriving at the right-hand submitting (giver) rapier (11), passes to the left in the rapier in the direction of head *A*. In addition, the air flowing around the nozzle (12) creates a suction stream in the nozzle (12) and immediately in front of the weft entrance (the air entering the rapier head (11) flows towards the tip of rapier tube *A* and, hence, the suction develops in the nozzle (12)).

Thus, in each rapier, the traction on the thread is created by a pair of air streams: soaking-up (suction) and forcing-on. In the left rapier, the greater part of the traction is created by a soaking-up stream, and an auxiliary one – a forcing stream – comes from the right rapier. In contrast, in the right submitting rapier, the effect of the soaking-up stream on the weft tractionis much less than that of the forcing stream. The rapiers move apart, by which time the weft has moved through the left rapier. When the rapiers have completed their outward movement, the new weft is beaten up and cut at the left-hand selvedge.

The tips of the rapiers have chamfers to move warp threads apart in the shed, and these tend to deflect the air streams negatively. The closest distance between the ends of rapiers when meeting in the shed is ℓ = 2–6 mm. The internal tube of the rapiers is rigidly held (d_2 = 5 mm) by rubber collars inside the external tube (diameter 16 mm). The rapiers are guided in the shed by the reed and the shed branches. Air is supplied at 0.03–0.06 Mpa pressure. At the left-hand rapier, the air pressure is 0.01 Mpa higher to create the soaking-up stream.

The planetary directing mechanism (Fig. 5.16(b)), with external gearing, provides continuous back-and-forth motion in a straight line of the hinges joining the rapier (9) to the rod (15) and the hinges joining the rapier (11) to the rod (15). The stationary solar wheel Z_3, using the rotating carrier (13), rolls two satellite gear wheels Z_4, Z_5, making planetary wheel Z_6 rotate. The carrier (13) rotates on axis O_1 by means of a drive Z_1, Z_2 (Fig. 5.16(a)) from the main shaft (14). The rapier (9) (Fig. 5.16(b)) with gear wheel Z_6 is controlled by the rod (15). The drive provides a high operating speed for the ATPR-120 weaving machine of 360 weft insertions per minute. Transfer of the end of the weft thread from the right rapier to the left can be made more reliable by:

- Increasing the diameter d_1 of the tubes of the receiving rapier (Fig. 5.16(c)).
- Improving the flow of compressed air to the left-hand rapier to improve weft suction (Fig. 5.16(d)).
- Increasing the transfer time of the end of the weft thread by applying rapiers of different sizes and making their movement in the shed compatible (Fig. 5.16(b)).

If rapiers are equal ($L_1 = L_2$), they meet briefly in the middle part of a shed when the angle of rotation of the main shaft of the weaving machine is 180°. If rapiers are of different sizes, the right-hand one (11) (L_2, longer) passes further before meeting with the left-hand one (9) (L_1) at 180°. Then, the right-hand rapier leaves the shed at proceeding accompanying movement of the left rapier by 12°–15° of main shaft of the weaving machine. Separation of the rapiers occurs at 192°–195°.

5.8 Weft insertion by an electromagnetic drive

Instead of a sley bar, the stator of a linear asynchronous motor providing a running magnetic field can be used to move a small-sized weft-picking device through the warp shed. The British firm Wilson & Longbottom developed a weaving machine with reed width of 2 m, and the Spanish firm Experiencias Industriales produced a design with a reed width of 5 m. These linear motors are powered by a three-phase electricity supply and the directing combs of the micropicking device, made from an aluminium alloy, achieve a speed of 25–32 m/s. Braking of the micropicking device occurs through an electromagnetic field. However, they have not become established in the industry. A basic problem with linear electric motors is the need for additional cooling of the stator.

5.9 Weft insertion by the inertial method

A thread can be accelerated by means of two contra-rotating cylinders having a nip (point or line along which they make contact). Through the action of inertia, a weft thread, depending on its rigidity, can pass a certain distance through an open shed. A constant speed of rotation of the cylinders tends to cause the thread to fold up as it is slowed down by air moving past the leading parts of the yarn. To keep the thread straight, it is necessary gradually to reduce the speed of its release from the cylinders in proportion to the air-induced braking action in the shed. A special drive is required for the purpose, causing the cylinders to rotate at a variable speed. A more successful variant for producing inertial movement of a thread uses two (contra-rotating) cones rotating at constant speed. The weft thread is moved by a guide between the nipline (the line of contact) of the cones to preserve its rectilinear form as it is picked into the shed to form a fabric. In this way, it is possible to insert wefts from flax, jute, etc. over a distance of 1m or so to create narrow fabrics.

5.10 Comparative analysis of different methods of weft insertion

The method of weft insertion by shuttle enables woven fabric with a highly uniform distribution of threads and high quality natural selvedges to be produced. Weft insertion by microshuttle, however, enables woven fabric of relatively reduced weft density to be made. The Inertial Small Picking Device (projectile) permits a rate of weft insertion to 20 m/s (the shuttle speed is up to 13 m/s). However, the artificial selvedges produced are of somewhat inferior quality (leno, tucked-in, stitching or others type of selvedges).

Weft insertion by rapiers requires extra time for the withdrawal of the rapiers from the shed before the weft can be beaten up. Rapid withdrawal of rapiers, so as to maintain a higher weaving speed, increases the loading on the rapier drive. Weft insertion by air and water jets does not suffer these limitations, and therefore these machine have a higher productivity rate. However, some twist loss can occur during weft insertion (due to the weft tip being free during weft insertion) which can cause some unevenness of structure in the woven fabric. The pneumo-rapier method of weft insertion has the potential to increase the reliability of weft insertion by preventing a loss of air pressure in rapiers during weft movement in a shed. A running magnetic field has the potential for a compact weft-picking device of low inertia and, hence, the potential for high weaving performance. But, so far, this has remained unrealized. The movement of weft solely by its own inertia has also remained a concept only.

5.11 Questions for self-assessment

1. What are the main features of the different methods of weft insertion used in weaving?
2. What are the advantages and disadvantages of shuttle weaving?
3. Using the typical shape of a velocity/time graph, explain the distinct areas of motion of the shuttle on a weaving machine.
4. What are the different versions of picking mechanisms found on shuttle weaving machines? Identify their advantages and disadvantages.
5. What are the features of a positive shuttle drive on a ribbon weaving machine?
6. How is the movement of shuttles on a circular weaving machines achieved?
7. What is a microshuttle?
8. Compare the different types of microshuttles, with their advantages and disadvantages.
9. What is an Inertial Small Picking Device (projectile) of weft?
10. What is a rapier with reference to a weaving machine?
11. What are the situations in which rapiers may cause interruptions to weft insertion on a shuttleless weaving machine?
12. What are the basic features of continuous weft insertion by rapiers on the Jentilini-Ripamonti weaving machine?
13. What are the basic features of the pneumatic (air-jet) method of weft insertion?
14. What devices other than the main nozzle are used for creating moving air in a pneumatic weaving machine?
15. What difference in air pressure is necessary for cotton weft threads as compared with silk threads, and what is the main reason for any difference?
16. How is an increased weaving width realized in the weft thread inserted on a pneumatic weaving machine?

17. What happens to a drop of water when a weft thread is picked in the shed of a water-jet weaving machine?
18. How many grammes of water should be used to insert one weft thread?
19. What are the advantages and disadvantages of air-jet and water-jet methods of weft insertion?
20. Explain the basic arrangement of components in the pneumo-rapier method of weft insertion.
21. What are the basic advantages and disadvantages of the pneumo-rapier method of weft insertion?
22. How can the reliability of the pneumo-rapier method be improved?
23. In what ways can an electromagnetic field be used as a means of weft insertion?
24. How can weft insertion in a weaving machine be achieved by utilizing the inertia of the weft itself?

5.12 References

1. Gordeev V.A. and Volkov P.V., 'Weaving', Leg. and Pitsh. Prom., Moscow, 1984 (in Russian).
2. Choogin V.V., Kahramanova L.F. and Nedovisiy M.N., 'Technology of Weaving Manufacture', State Technical University, Kherson, 2008 (in Russian).

6
Woven fabric formation: principles and methods

DOI: 10.1533/9780857097859.109

Abstract: This chapter describes woven fabric formation on shuttle weaving machines and shuttleless weaving machines. The peculiarities of the conditions of woven fabric formation by the different mechanisms employed are presented: the methods of moving the weft into the fell of the woven fabric, and methods of weft beat-up. A method of calculation of parameters of woven fabric formation for using in practice is given. The unique properties of the elastic system of fabric formation (ESFF) are discussed, and the chapter closes by covering the methods of reducing tension in the ESFF.

Key words: woven fabric formation, front-centre beating-up, point beating-up, elastic system of fabric formation (ESFF).

6.1 Introduction: woven fabric formation

The process of forming a length of woven fabric consists of two operations: inserting the weft thread into the cloth fell and the formation of fabric cells. In practice, two methods of moving weft thread into the cloth fell are used: 'full-width' and 'point' (Ref. 1 and Ref. 2). If the 'full-width' method is used, the weft is combines into the cloth fell simultaneously across the full-width of the woven fabric. With the point beat-up method, the weft is combined into the cloth fell section by section (a short section of weft at a time). The formation of an element of fabric is carried out in three different ways: by beating-up, condensing, or embedding.

Beating-up is the method of forming a woven fabric by the rapid application of force by the reed on the newly inserted weft to move it into the cloth fell. When weft is moved into the cloth fell by the *condensing* method, the fabric-forming operation only involves crossing the warp threads; the formation of an element of woven fabric occurs by means of the shifting of the new weft into the cloth fell by the crossing action of the warp threads. The *embedding* method involves the continuous operation of formation of the fabric element by pressing the blades of a reed (or laying needles, sliding disks, etc.) on the weft until the moment it is positioned in the cloth fell following the crossing of warp threads in the course of formation of the new shed. The movement of the weft thread into the cloth fell by the 'full-width' method, following completion of the picking operation, is carried out by the reed, situated on the sley mechanism.

The sley mechanism consists of a beam representing the base or carrier which carries the fabric-forming component (the reed), which may also be used for guiding the weft insertion device (Ref. 1 and Ref. 2). On shuttle weaving machines, the four-bar sley mechanism is widely used, with sley movement obtained by means of the Schönherr crank (Fig. 6.1(a)). The crank (1) (R), by means of a connecting rod (2) (L), gives a reciprocating circular movement to the sley beam (3) with the reed (4) mounted on it. The race (5) of the sley beam (3) serves as one of the guides of the shuttle (6). There is no *dwell* period (i.e. duration in which the sley stays without moving) during the cycle of the sley and reed. In producing dense furnishing fabrics, weft thread undergoes double beating-up (Beridot, Fig. 6.1(b)). The trailer hinge of the crank (1), passing positions M_1 and M_3, establishes links (2 and 3) on one line, and the reed (4) at the extreme forward position of beating-up.

For producing fabrics with a looped pile (Reh, Diederichs) (Fig. 6.1(c), (d), (e)) a link (2) is added into the drive of the sley beam (1) to increase the shorter length OO_1 to the longer length OO_2. At setting OO_1, the reed (3) does not move up to the fabric fell ($l_1<l_2$) for the full length of connecting rod L; therefore, the weft threads move to positions P_1 and P_2 (l, a 'light' beating-up) to form the pile threads (4). The reed (3) (Fig. 6.1(c)) will occupy its extreme forward position at the maximum distance $OO_2 = R + l_2$ (Fig. 6.1(d)), resulting in a 'heavy' beating-up). On the third or fourth (depending on the type of weaving interlacing) 'heavy' beating-up P_3 (Fig. 6.1(e)), the weft (5) makes a final securing of the threads (4) and simultaneously displaces the previous two weft threads to the cloth fell. The threads (4) form a loop (6).

Movement of the link (2) from l_1 to l_2 is made by a tappet (7) by means of a lever (8) and tension rod (9). Repositioning of the link (2) from l_2 to l_1 is made by a spring by the axle O_1 or O_2 (this spring is not shown in Fig. 6.1(c), (d)). On shuttleless weaving machines, the sley undergoes a longer dwell phase (for an angle of rotation of the main shaft of the weaving machine by 150°, ... , 250°) to afford optimum working conditions for the weft insertion mechanism. The reed drive with coupled (conjugate) tappets has been most widely adopted. In Fig. 6.2(a), the coupled (conjugate) tappets (1, 2) which control the lever (3) drive the sley beam (4) together with the reed (5) and guiding plates (6) for the weft picking device (7) mounted on it. The profile of the tappets (1, 2) os designed to

Woven fabric formation: principles and methods 111

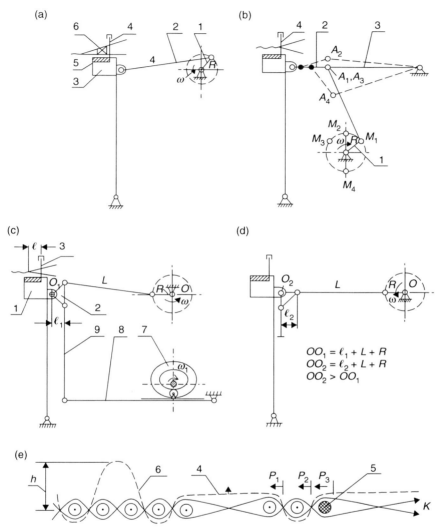

6.1 Crank mechanisms for reed drive. (a) Schönherr; (b) Beridot; (c), (d) with compensator of crank arm length; (e) formation of an obverse loop in a fabric on (c) and (d). Refer to text for detailed explanation of components.

suit the timing required by the design of the weft picking mechanism. On the pneumatic weaving machine (Elitex), a cam (1) (Fig. 6.2(b)), operates the sley beam (2) by means of a rod (3), lever (4), link (5) and sley sword (6). The dimensions of all the links and the arrangement of hinges provide an effective sley dwell period of 90°, … , 130°.

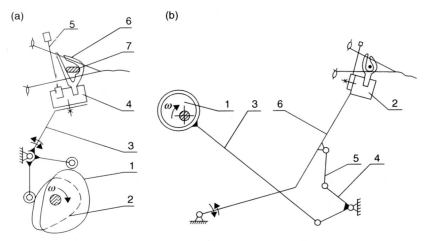

6.2 Tappet type sley mechanisms of shuttleless weaving machines. (a) Sulzer; (b) Kovo.

6.2 Fabric-forming mechanisms

To form a length of woven fabric, devices such as the reed, pressure plates, pressure disks, crown disks, sliding concentrators etc. are used (Ref. 2). The highest quality of woven fabric and the greatest degree of weft consolidation in the cloth fell is achieved by the *full-width* application of the reed. This is when all the warp threads are uniformly distributed across the width of the fabric and all of the reed wires or blades operate simultaneously along the length of weft.

A high weft density of fabric can be achieved by use of beating-up plates (Titov, Fig. 6.3(a)), which consist of separate profiled plates (1) set on a shaft (2). The beating-up of the weft thread (3) is carried out according to the 'full-width' method. A reduced density of the woven fabric is obtained on the Jentilini-Ripamonti machine (Fig. 6.3(b)). Teeth (1) on fabric-forming disks (2) gradually displace the weft (3) in the crossed shed into the cloth fell (4). The tip A of the tooth (1) gradually goes under the cloth fell (4) and the base plate (5). Sliding of the profile of tooth (1) across the weft thread (3) causes some abrasion of the weft thread. The incorporation of the weft thread into the cloth fell is carried out by the full-width method.

A very simple method of *full-width* incorporation of the weft into the cloth fell is presented in Fig. 6.3(c). Two shafts (1, 2) with flexible needles (3) form brushes which, on counter-rotation, displace the weft (4) into the cloth fell and form a new fabric cell. Any lack of uniformity of needle operation between warp threads and the compliance of needles may cause a woven fabric of reduced density.

Woven fabric formation: principles and methods 113

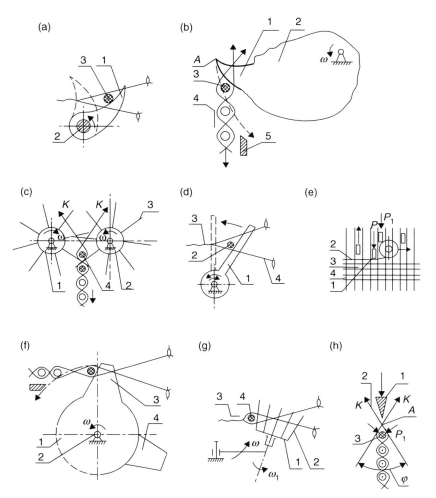

6.3 Fabric-forming elements. (a) Beating-up comb (D.B. Titov); (b) a sliding comb (Gentilini-Ripamonti); (c) brushes; (d), (e) plates (D.I. Popov, Rossman); (f) fabric-forming disks (E.A. Onikov, Pignone-Smit); (g) a crown (Sagem, Fayole-Ancet); (h) concentrator (S.A. Dynnik).

The point beat-up method of weft incorporation into the cloth fell on multished weaving machines is achieved by local pressure applied on the weft by vibrating plates, toothed disks etc. Thin plates (1) (Cerdans, Rossman, Popov, Fig. 6.3(d) and (e)) press the weft (2) into the cloth fell (3) over a small section comprising between two and six warp threads (4). When the plates (1) are withdrawn, the loss of pressure may cause the weft (2) to slip back to some extent from the fabric elements previously formed. This precludes the production of a dense fabric by means of the point beat-up of weft.

On multished machines, fabric-forming disks (1) (TMM, Pignone-Smit, (Fig. 6.3(f)) with a spiral arrangements of teeth (3, 4) are mounted on the shaft (2). The first tooth (3) carries out the initial incorporation of an element of weft thread into the cloth fell, and the second tooth (4) completes the formation of that element. On the circular weaving machine (Fayolle–Ancet), the point beat-up of the weft into the cloth fell is carried out by a crown disk (1) (Fig. 6.3(g)) with needles (2). The crown (1) rolls along the circular fell (3), and needles (2) serially pass between the warp threads and push weft (4) to the cloth fell (3).

Formation of fabric by the condensing method is performed by a circular weaving machine (Dynnik S.D. TRD-425 (Fig. 6.3(h) (105.h)). The concentrator (1) slides along the cloth fell and presses on the point at which the warp threads (2) cross. A combination of two factors (the tension of threads K, and forces of pressure P_1 of the warp threads) cause crossing point A to be displaced in the direction of the fabric already formed. Thus, the angle φ will increase, and on the application of pressure P_1, the warp threads will displace the weft (3) towards the previously beaten-up weft. However, use of the point beat-up method precludes the formation of a fabric of high density.

6.3 Formation of the woven fabric cell

A fabric cell is the extent of a set of warp threads associated with an element of one weft thread (Ref. 2). With the 'full-width' method of inserting the weft thread, the part of the weft in the fabric element is equal to the width of the cloth-fell. With the point beat-up method, only that part of weft which is subject to the local pressure of the fabric-forming elements (plates, disks, etc.) is affected. In practice, a larger assortment of fabrics is developed with 'levelling' by the method of beat-up. To better explain this concept, we will consider two more possible variants of the beating-up of weft into the cloth fell: without levelling, and with levelling.

The beating-up of the weft in an open shed (Fig. 6.4(a)) (without levelling) allows the attachment of weft (1) by reed (2) against a reduced resistance of forces P_1 from the warp threads (3). However, after reed (2) has moved away, the weft (1) is likely to be pushed away from the cloth-fell by forces P_1 as the weft is not secured by the warp threads. This occurs owing to the considerable difference in the rate of movement of the reed and the harness. Therefore, achieving a fabric of average density is not possible.

To develop an open 'gauze' fabric, the beating-up of the weft thread is carried out in the level phase (Fig. 6.4(b)), when the compression forces P_2 from the warp prevent the weft thread (1) from easing back from the

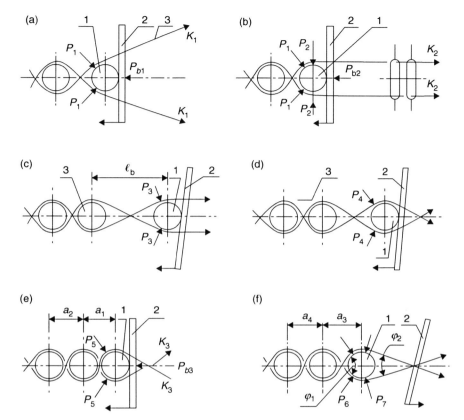

6.4 The 'full-width' beating-up of weft. (a) Beating-up in an open shed; (b) levelling; (c) size of levelling; (d) moving weft thread in a closed cell; (e) beating-up in the closed cell; (f) withdrawal phase of reed.

cloth-fell after the withdrawal of the reed (2). However, the same forces P_2 act against the beating-up of the weft thread ($P_{b2} > P_{b1}$). Both variants develop a fabric 'without levelling'.

To prevent weft 'slipping back' from the cloth fell following the withdrawal of the reed, it is necessary to restrain the weft by forming a new shed with the warp threads. To allow time for the warp threads to form a shed behind the newly inserted weft thread on the withdrawal of the reed, a level phase (when the warp threads go beyond the average height of their working range) must be obtained before the beating-up process. This is the development of a fabric 'with levelling'.

Figure 6.4 presents the basic phases of formation of a woven fabric with levelling under '*the condition of equality*' of tension K_i of the top and bottom branches of a shed: levelling (Fig. 6.4(c)), weft movement in the closed cell by reed (Fig. 6.4(d)), the extreme forward position of the reed during the beating-up phase (Fig. 6.4(e)), and the phase of reed withdrawal from the cloth fell (Fig. 6.4(f)). The distance ℓ_b (Fig. 6.4c) between the reed (2) and the cloth fell (3) during the 'moment of levelling' (at which the formation of a new shed begins) is called the 'size' of levelling; it is measured either as a linear distance (mm), or as an angular position (degrees of rotation α_{ms}) of the main shaft of the weaving machine. In practice, the size of levelling is established within $\ell_b = 20, \ldots, 100$ mm at the time of rotation of the main shaft of the weaving machine $\alpha_{ms} = 10°, \ldots, 60°$.

The further movement of the weft thread (1) (Fig. 6.4(d)) by the reed (2) to the cloth fell (3) is carried as a closed cell. An increase in the angle of interlacing of the weft thread (1) by the warp threads considerably increases the forces of friction between the warp thread and the weft. To overcome this, it is necessary to increase the tension of the warp threads. In the extreme forward position of the reed (2) (Fig. 6.4(e)), the weft (1) is exposed to maximum compression on the warp threads P_5 at the maximum values of tension of the warp K_3 and force of beating-up P_{b3}. In this situation, following withdrawal of the reed (2) (Fig. 6.4(f)), the slipping back of the weft thread (1) from the cloth fell under the influence of forces P_6 is limited by the opposing forces P_7. Owing to the considerable difference in the angles of crossing of the warp threads before the weft φ_1 and after it φ_2 ($\varphi_1 >> \varphi_2$), there is a withdrawal (slippage) of the beaten-up weft thread away from the previously beaten weft threads (distance a_1 changes to a_3). Thus, for weaving fabrics of higher weft density by the beating-up operation, it is necessary to increase the intensity of conditions for fabric formation. Choogin has proposed a method by which the weft thread can be incorporated into the open angle of the shed with a reduced resistance from warp threads while preventing the weft thread from withdrawing from the cloth fell. This is achieved by prolonged contact of the reed with the weft before its being secured in the fell by crossing the warp threads to create a new shed. This method requires the modification of the profile of the coupled tappets (1, 2) (Fig. 6.2(a)).

6.4 Parameters of woven fabric formation

At the cloth fell, there may be slippage not only of the last beaten-up weft thread, but also some of the preceding weft threads. This is a characteristic which is part of the concept of the 'fabric formation zone' in which the weft threads involved in the beating-up phase may slide back along the warp threads at the cloth fell (Ref. 1 and Ref. 2). The extent of the fabric formation zone is characterized by the number of displaced wefts (2, ..., 10 weft threads) and depends on the type of fibre involved, the structures of threads and the woven structure concerned, as well as the conditions of warp mounting on

Woven fabric formation: principles and methods   117

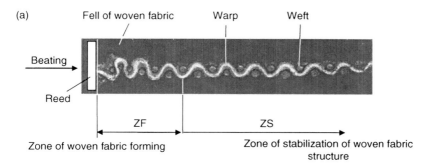

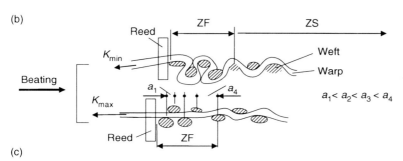

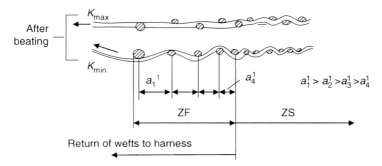

6.5 Microsection of a woven fabric along warp threads under variable tension of shed branches. (a) Cotton fabric of plain weave at the extreme forward position of reed; (b) linen fabric of plain weave at the extreme forward position of a reed; (c) linen fabric of plain weave after the withdrawal of the reed from the cloth fell.

the weaving machine. Figure 6.5 shows microsections of the cloth fell of a plain-weave fabric woven from cotton and flax yarns obtained under different levels of tension, together with the bottom and top branches of the shed.

Vasilchenko (Ref. 3) has presented the zone of formation (ZF) in a photograph (Fig. 6.5(a)) of the beating-up phase of the weft when producing a plain-weave fabric from cotton threads. The reed presses on the cloth fell and creates conditions for considerable bending of the warp threads of the shed branches, reducing their tension. The density of the arrangement of seven weft threads in the fabric formation zone has increased (ZF = 7 weft yarns), following which there is a zone of stable wefts (ZS).

In Fig. 6.5(b) and (c), Parfenov (Ref. 3) presents a sketch of the ZF of a plain-weave fabric woven from linen threads during two phases: at the point of beating-up, and at the withdrawal of the reed (Fig. 6.5(b)). The size of the ZF of the fabric is four weft threads. Here, it is necessary to pay attention to the essential circumstances: at beating-up (Fig. 6.5(b)) in the ZF we observe *the inequality* of distances between weft threads as $a_1<a_2<a_3<a_4$; at reed withdrawal, the opposite correlation of these distances $a_1^1 > a_2^1 > a_3^1 > a_4^1$ occurs. Such displacement of wefts occurs on each beating-up.

Owing to the existence of zone of fabric formation, ZF, gradual variation of the tension of warp threads during the weaving cycle, differences in the elastic properties of warp threads and the fabric in the ESFF, and the differences of the phases of operation of the basic working mechanisms of a weaving machine, the reed meets the cloth fell before reaching its extreme forward position. This leads to what is termed 'weft stripe' WS, which is the displacement of the cloth fell by the reed during its movement to its extreme forward position (front dead centre).

The size ℓ_u of weft stripe (or 'cloth fell length') on beating-up depends on many factors. It decreases with an increase in tension of the warp threads, and the stiffness and the frictional properties of threads. The weft stripe increases in size when there is an increase in the density of the weft threads in the fabric, intensity of interlacing, smoothness of threads, resistance to compression of the weft, and the volume fraction of yarn in the fabric. In practice, weft stripe is found to be 2, ... , 10 mm. An excessive length of weft stripe leads to increased wear of warp threads by droppers, by heald eyes, and by reed wires or teeth. Positive effects of the weft stripe on the weaving machine are:

- the weft stripe compensates for sharp changes or fluctuation in ESFF tension;
- it creates conditions for autoregulation of the mode of release and the tension of warp threads in the resilient system of fabric formation.

Autoregulation of the ESFF of a weaving machine can be described as follows. In the event that the tension changes in the ESFF (e.g. if the tension increases beyond its set parameters), the size of the weft stripe decreases. As soon as the weft stripe decreases in length, then there will also be a decrease in the deformation of stretching of the warp threads in the beating-up phase. Reduction of the deformation of stretching of the warp threads will lead to

Woven fabric formation: principles and methods 119

a reduction in the level of warp tension in the ESFF. Thus, there is a gradual cause-and-effect process of self-alignment (autoregulation) of tension in the ESFF *without* the intervention of the operator of the weaving machine. When adjusting the weaving machine, the size of the weft stripe can serve as *criterion of the optimality* of its mounting (drawing-in) parameters for the development of a given structure of woven fabric.

The tension of warp threads K increases in the phase of formation (beat-up phase) of a length of fabric to a maximum value, with the simultaneous fall in fabric tension of K_T. The parameter P_{bi}, the force of beating-up, is characterized by a difference of tension of the warp threads and in the fabric at the extreme forward position of reed (N/thread). Gordeev (Ref. 1) gives a simple expression of the force of beating-up involving the size of the weft stripe ℓ_u and factors of rigidity of the warp threads and the woven fabric in the ESFF of the weaving machine:

$$P_{bi} = K - K_T = \ell_u (C_0 + C_T). \qquad [6.1]$$

where C_0, C_T = stiffness coefficients of warp threads and fabric, N/mm (see Chapter 1, Section 1.4).

For example, for $\ell_u = 5$ mm; $C_0 = 100$ N/mm; $C_T = 150$ N/mm,

we obtain: $P_{bi} = 5(100 + 150) = 1250$ N.

According to Kutepov (Ref. 2), a complex indicator of a woven fabric structure, on which the complexity of the process of formation the length depends, is the factor of volume filling of the fabric by yarns: H_v. The indicator considers the thickness and the type of fibre of the threads, the density of warp and weft in the fabric, the connectivity of the interlacing of threads in the fabric (characterized indirectly by shrinkage of the threads), and the thickness of the fabric. The magnitude of factor H_v is given by the following equation:

$$H_v = 7.85 \cdot 10^{-6} \cdot [T_o \cdot P_o \cdot C_{fo}^2 \cdot (1 + 0.01 \cdot A_o) + T_u \cdot P_u \cdot C_{fu}^2 \cdot (1 + 0.01 \cdot A_u)] / B_c. \qquad [6.2]$$

where T_o, T_u = linear density, g/km, tex;

B_c = thickness of the woven fabric, mm;

$$B_c = 0.0316 \cdot \left(C_{fo} \cdot \sqrt{T_o} + C_{fu} \cdot \sqrt{T_u} \right), \qquad [6.3]$$

A_o, A_u = the crimp of the warp and weft threads, %;
P_o, P_u = the density of the warp and the weft, threads/dm;

C_{fo}, C_{fu} = the factors depending on the kind of fibre of warp and weft threads.

(For: cotton, $C_f = 1.25$; wool, $C_f = 1.33$; worsted spun, $C_f = 1.28$; flax, $C_f = 1.23$; artificial silk, $C_f = 1.39$; nylon, $C_f = 1.35$; polyester, $C_f = 1.17$; etc.)
For example: for cotton fabric, $T_o = 25$ tex; $T_u = 30$ tex; $P_o = 200$ threads/dm;

$$P_u = 180 \text{ threads/dm}; A_o = 9\%; A_u = 7\%; C_{fo} = C_{fu} = 1.25;$$

we obtain: $B_c = 0.0316 \cdot \left(1.25 \cdot \sqrt{25} + 1.25 \cdot \sqrt{30}\right) = 0.41 \text{ mm}$

$$H_v = \frac{7.85 \cdot 10^{-6} \cdot [25 \cdot 200 \cdot 1.25^2 \cdot (1+0.01 \cdot 9) + 30 \cdot 180 \cdot 1.25^2 \cdot (1+0.01 \cdot 7)]}{0.41} = 0.336$$

From experience (Choogin):

light woven fabrics, $H_v = 0.10 \div 0.29$;
woven fabric by mean value of close, $H_v = 0.30 \div 0.49$;
closely woven fabric, $H_v = 0.50 \div 0.90$.

Depending on size H_v, one can determine (Ref. 2) the tension of warp threads at beating-up with the following equation (cN/thread):

$$K_{bi} = C_1 \cdot H_v^3 + \frac{0.3}{H_v} + C_2 \qquad [6.4]$$

where C_1, C_2 = the factors relating to a particular fibre; for:

cotton, $C_1 = 478$, $C_2 = 12$;
wool, $C_1 = 450$, $C_2 = 15$;
flax, $C_1 = 520$, $C_2 = 17$;
silks, $C_1 = 400$, $C_2 = 9$.

For example, for cotton close-woven fabric $H_v = 0.52$; $C_1 = 478$; $C_2 = 12$, we obtain the tension of warp threads at beating-up:
$K_{bi} = 478 * 0.52^3 + 0.3/0.52 + 12 = 79.8$ cN/thread.
In connection with the repeated influence of working on a warp thread, Borodovsky (Ref. 2) defined the permissible deformation of stretching of warp threads within tolerance $[\delta]$, (%):

$$[\delta] = \frac{102 \cdot P}{(T_o \cdot E_{tc})}, \quad [6.5]$$

where E_{tc} = the linear-cyclic modulus of the warp thread, m; for cotton thread, E_{tc} = 400, ... ,480 m; for woollen thread, E_{tc} = 50, ... ,100 m. For example, for P = 1250 N; T_o = 25 tex; E_{tc} = 400 m; we obtain:

$$[\delta] = \frac{102 \cdot 1250}{(25 \cdot 400)} = 12.67\%.$$

From this equation, it is possible to determine a force of beating-up (P) which does not affect the integrity of threads on repeated loadings.

6.5 Ring temples

In the phase of full-width beating-up, the weft thread joins the cloth fell in the form of a *straight* line; thus, the length of cloth fell and the inserted weft are defined by the width of drawing-in of warp threads in the reed (Ref. 1). After the withdrawal of the sley to its back position, the warp threads bend (crimp) the weft according to the parameters of the structure of the fabric and the physical properties of the threads. As a result, the cloth fell reduces in width (contraction of fabric width) and, correspondingly, the warp threads settle down with reduced spacing between them. At the subsequent beating-up, as the reed moves up to the beat-up position, it stretches the cloth fell. Thus, the warp threads in the selvedges experience wear due to a considerable *bending* force applied by the reed wires. Breaking of selvedge threads would increase as a result. To prevent this, special auxiliary devices (temples) are used on weaving machines. Temples counteract the shrinkage (reduction of length) of the cloth fell by stretching the fabric at the cloth fell to the same width as that of the reed. There are various designs of temple according to the fibres being woven.

Ring temples with needles are the most widely used temples (Fig. 6.6(a)). On the axle (1) is an eccentric bush (2), on whose rings (3) needles (4) type set without clamping and latching. All bushes (2) are fixed to the axis (1) by the head (5) of the axis and an inclined washer (6), usually by means of a nut. In the beating-up phase (Ref. 2), the reed (7) (the Fig. 6.5(b)) moves the weft (8), together with the cloth fell, to the temple cap (9) at a distance of about 2 mm (this value is optimal). Thus, the fabric (10) is held within the edges of the cap (9) and is pierced by the needles (4) on the rings (3).

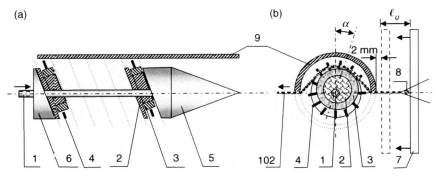

6.6 Ring temples. (a) Arrangement of rings with needles on axis; (b) route of fabric around the edges of a temple cover.

The fabric releases from the needles (4), over the front edge of the cover (9) and moves to the breast beam on the weaving machine. The inclined position of the rings (3) with needles (4) (Fig. 6.5(b)) stretches the cloth fell to the drawing-in width of the warp threads in the reed (7). In this way, damage to the extreme threads in the selvedges is decreased. From experience, it is known that the optimum angle of inclination of the maximum radius of the eccentric bush corresponds to $\alpha = 15°$. For the manufacture of silk fabrics, temples with corrugated or rubber platens without needles are used; for thin fabrics from cotton, wool, etc., rough platens with a small notch are used.

6.6 Methods of easing of fabric formation

The most effective way of reducing the strain in the threads in the fabric formation phase is the creation of a *variable tension* (also termed 'unbalanced' or 'unequal') *of the bottom and top branches* of the shed. Under variable tension, the weft moves, in response to pressure from the reed, along the more highly tensioned branch and bends the warp threads of the more lightly tensioned branch with greater ease. As a result, the resistance to formation of a fabric length decreases and, hence, the necessary level of force of beating-up decreases. As a result, for the formation of a fabric of a given density, a reduced level of the drawing-in tension is required.

Where there is variable tension in the branches of the shed, there is the danger of the cloth fell sliding along the height of reed during the beating-up phase. Therefore, it is necessary to establish the optimum angle of beating-up (i.e. the angle between the plane of the reed and the fabric at which there is no slippage of the cloth-fell). Parfenov considers that the optimum value of the angle of beating-up at which the force of beating-up P_{bi} is directed along the bisector, enclosed between the fabric and the continuation of the plane of the more lightly tensioned branch.

6.7 Comparative analysis of the methods of fabric forming

The method of moving the weft thread to the cloth fell simultaneously across the full-width of the woven fabric enables the formation of a stable fabric structure. The 'point beat-up method' joins the weft to the cloth fell section by section; therefore, the woven fabric may form a structure of reduced stability. Rapid beat-up of weft into the cloth fell by the reed forms a stable fabric. In the 'condensing' method, the weft is moved into the cloth-fell by the action of the crossed *warp* threads; however, this method precludes the formation of a closely woven fabric.

Weft 'embedding' has advantages in comparison with the beating-up and condensing methods as it supports the continuous formation of fabric by pressing the reed on the weft until it is positioned in the cloth fell, following the crossing of warp threads in the course of formation of a new shed, ensuring formation of a closely woven fabric with reduced forces. A reed drive with coupled tappets permits a high degree of flexibility with regard to reed movement, unlike a crank mechanism, the reed movement of which is fixed.

When compared with the use of pressure plates or pressure disks, crown disks, sliding concentrators etc., the best quality length of woven fabric is achieved with the use of a reed, which employs the *full-width* method of beat-up action on the weft. Autoregulation of the ESFF on a weaving machine is the established method of ESFF tension control without the intervention of the weaving machine operator, and achieves a stable fabric structure. The use of variable tension in the bottom and top branches of the shed, combined with an optimum value of the angle of beating-up, provides for reduced strain in the warp threads during the fabric formation phase.

6.8 Questions for self-assessment

1. What is the difference between the 'full-width' and 'point' methods of incorporating weft into cloth fell?
2. Explain what is meant by 'beating-up', 'condensing' and 'embedding' of weft in the cloth fell?
3. What are the different types of sley mechanism used in shuttle weaving machines?
4. How is a loop of warp formed in a fabric?
5. What are the differences in the sley mechanisms of the Sulzer and Kovo weaving machines?

6. What are the devices used to supply weft to the cloth fell?
7. What is a 'cell of woven fabric'?
8. What are the stages of supplying weft into the cloth fell in the full-width method of beating-up?
9. What are the disadvantages of forming a woven fabric 'without levelling' (beating-up with an open shed) and 'in a level phase' (beating-up with a closed shed)?
10. Why is it necessary to have 'fabric formation with levelling' (i.e. beat-up after a change of shed) with the majority of different of structures of woven fabric?
11. Explain the stages of woven fabric formation with shed levelling.
12. What is the 'beat-up weft stripe'? What are the causes of its formation? On what does its size depend? How does it influence the elastic system of fabric formation (ESFF)?
13. What is 'autoregulation of the elastic system of fabric formation'?
14. What is the 'force of beating-up' and how is it possible to define its size?
15. How is it possible to define the magnitude of deformation of stretching of warp threads at the beating-up of a weft thread?
16. What is the purpose of temples? How do they improve the beating-up of weft?
17. How is it possible to reduce the tension (strain, stress) of warp threads at fabric formation?
18. What is 'the optimum angle at beating-up of a weft thread'?

6.9 References

1. Gordeev V.A. and Volkov P.V., 'Weaving', 'Leg. and Pitsh. Prom.', Moscow, 1984 (in Russian).
2. Choogin V.V., Kahramanova L.F. and Nedovisiy M.N., 'Technology of Weaving Manufacture', State Technical University, Kherson, 2008 (in Russian).
3. Chepelyuk E.V. and Choogin V.V., 'Weft friction in weaving machines', National Technical University, Kherson, 2008 (in Russian).

7
Mechanisms for woven fabric take-up

DOI: 10.1533/9780857097859.125

Abstract: The features of fabric take-up from the working area and its winding onto the cloth beam are discussed in this chapter. Different types of take-up motion regulators allow the formation of woven fabric with various structural arrangements using weft of varying longitudinal thickness.

Key words: cloth regulator, woven fabric take-up, winding woven fabric onto the cloth beam.

7.1 Introduction: mechanisms for woven fabric take-up

Normally, the decision of a consumer buying a woven fabric is influenced by its appearance and its feel. The buyer is usually not interested in how the fabric is produced, or the actual arrangement of the yarns in the fabric. In Fig. 7.1(a), the arrangement of weft yarns with a *uniform* distribution is presented, where spacing a_i = Constant and thread density $P_u = 1/a_i$ = Constant (Ref. 1 and Ref. 2). However, if the diameter of the weft threads d_i is not constant, then the spaces between the the weft yarns b_i will also not be constant (Fig. 7.1(b)). Therefore, this arrangement of weft yarns is not appropriate for weft yarns of non-uniform diameter.

With an arrangement of weft yarns from uniform beating-up (Fig. 7.1(c)) gaps b_i between adjacent weft yarns are constant (b_i = const), but for a yarn with a high degree of irregularity in diameter ($d_i \neq$ const) the weft density will be be variable, thus: $P_u = 1/a_i \neq$ const. Such a fabric will appear to have a *uniform* density of threads. Such an arrangement of weft yarns can be applied to fabrics that do not have a highly distinctive pattern of interlacing. For development of a fabric from a contoured yarn, such as those from threads with special effects (shiny, metallic), a special plan is needed for the distribution of weft threads, and both parameters will be variable, thus: $a_i \neq$ const; $b_i \neq$ const.

In practice, the actual structure of a woven fabric does not always display a high degree of uniformity. This occurs because the density of a fabric

126　Mechanisms of flat weaving technology

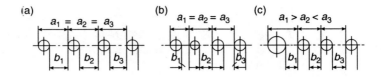

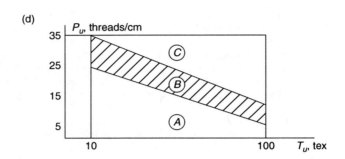

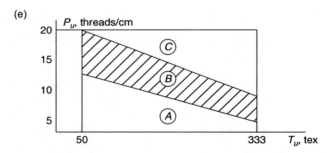

7.1 Distribution of the weft in a woven fabric: (a) and (b) uniform; (c) with equal intervals; (d) and (e) borders of zones of distribution of the weft.

depends on the weft thread density P_u and the thickness (or linear density, tex) T_u of threads. For example, in a higher-density fabric with closely woven weft, the distance between the centres of the weft yarns will be defined by their diameters (rather than the rate of fabric take-up). Therefore, it is practically impossible to produce a fabric with a uniform distribution of weft. Figure 7.1(d) shows the possible zones where a fabric with a uniform distribution of weft yarns may be produced, according to the experimental data of R.I.Harechko (Ref. 2):

- Zone A – at data P_u and T_u it is possible to have a fabric with uniform distribution of weft (a_i = const).

- Zone C – data P_u and T_u do not allow the formation of a fabric with a uniform distribution of weft.
- Zone B – transition zone, when the distribution of weft in the fabric obtained could be either uniform or non-uniform.

On the other hand, a fabric with a uniform beating-up of the weft (b_i = Const) can only be obtained with high-density weft (Fig. 7.1(e)):

- Zone A – for the given count of weft T_u and a low-density P_u it is impossible to obtain a fabric with uniform beating-up.
- Zone C – values of T_u and P_u allow a fabric with a uniform beating-up to be produced.
- Zone B – produces a transition-type fabric.

Woven fabric is moved away from the fabric formation zone by the take-up motion (cloth regulator). The purpose of the take-up motion is:

- to create the required density of weft in the fabric;
- to take up the fabric from the fabric formation zone;
- to wind the fabric that has been woven onto the cloth roller.

The weft density is set by the rate of removal of the fabric from the fabric formation zone by the take-up roller (emery roller). For best contact and to prevent fabric slippage, the emery roller is covered with special metal ('grater') or rubber tape with a corrugated surface. The action of the cloth beam regulators can be divided into types according to the method of fabric removal:

- independent of the fabric tension (positive);
- dependent on fabric tension (negative);
- programmed mechanisms.

Depending on the movement of the working element – the emery roller – and the corresponding fabric types, cloth beam regulators can be:

- continuous, with a constant or variable speed of fabric take-up; or
- periodic, with constant or variable length of fabric take-up.

Figure 7.2(a) shows a positive fabric take-up mechanism, which acts independently of the fabric tension and with a periodic uniform movement of the emery roller (2), as used on automatic shuttle weaving machines. The

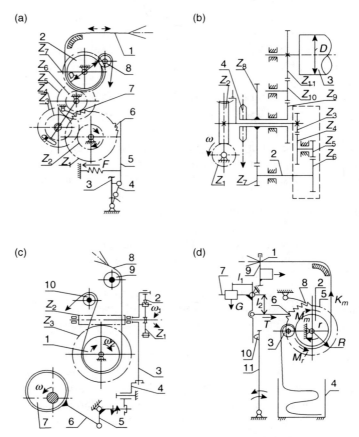

7.2 Mechanisms of the fabric take-up with different actions. (a) Discrete operation from a ratchet (Keighley); (b) continuous operation from a worm (Sulzer); (c) discrete operation from the crown wheel, ratchet and worm (Kovo) (d) dependent on the fabric tension (Schönherr). Refer to text for detailed explanation of components.

mechanism shown, of the Keighley type, is driven by the sley (3) using the ratchet mechanism (5, 6, 7, Z_1) and a geared reducer Z_2 to Z_7.

A hinged lever (5) is mounted on the sley sword (3) by means of an arm (4). When the sley moves to its back position, the arm (5) turns the ratchet wheel Z_1 by a pawl (6), moving it by one or two teeth. The three-stage cylindrical reducer which consists of interchangeable gear wheel Z_2, replaceable gear Z_3 and the four gear wheels Z_4 to Z_7, transfers this rotation to the emery roller (2). When the sley moves to its forward position, the moving pawl (6) carried on lever (5) moves to the next tooth of the ratchet. In this way, the fabric (1) is periodically taken up from the fabric formation zone. After passing the emery roller (2) and the platen (8), the fabric is wound up onto the cloth roller, which is turned by a special friction gear mechanism. This type of cloth regulator enables a fabric with a chosen level of weft density to be produced, which is selected by means of replaceable gear wheels Z_2 (15 or 30 teeth) and Z_3 (15–68 teeth). The density of the weft is given by the formula, in threads/cm:

$$P_u = \frac{1}{L_1} = C_i \cdot \frac{Z_3}{(m_x \cdot Z_2)}, \qquad [7.1]$$

where C_i = the constant factor of the regulator (C_i = 15 on AT weaving machines);

m_x = the number of teeth by which the ratchet (1, 2) turns;
L_1 = the length of fabric in cm which is taken away by the emery roller for a single turn of the main shaft of the cloth regulator:

$$L_1 = \pi \cdot D_c \cdot m_x \cdot \frac{Z_2 \cdot Z_4 \cdot Z_6}{(Z_1 \cdot Z_3 \cdot Z_5 \cdot Z_7)}, \qquad [7.2]$$

where D_c = diameter of the emery roller in cm.
For example, for $C_i = 15$; $m_x = 1$; $Z_1 = 108$; $Z_2 = 15$; $Z_3 = 20$; $Z_4 = 22$; $Z_5 = 72$; $Z_6 = 24$; $Z_7 = 35$; $D_c = 11.3$ cm; we obtain:

$$L_1 = \frac{\pi \cdot 11.3 \cdot 1 \cdot 15 \cdot 22 \cdot 24}{(108 \cdot 20 \cdot 72 \cdot 35)} = 0.0516 \text{ cm}$$

$$P_u = \frac{1}{0.0516} = 19.38 \text{ threads/cm}.$$

A positive fabric take-up mechanism which has a continuous speed of independent action, as found on the Sulzer shuttleless weaving machine, is shown in Fig. 7.2(b). The worm pair Z_1, Z_2 by means of the shaft (1), transfers rotation to four interchangeable gear wheels Z_3, …, Z_6. These gears turn gear wheels Z_7 to Z_{11}, via shaft (2), driving the emery roller (3) in a constant rotation. From the sprocket (4), movement is transferred via a chain to the fabric winding mechanism on the cloth roller. The weft density in the fabric P_u is changed by a set of interchangeable gear wheels:

$$P_u = \frac{C_z \cdot Z_4 \cdot Z_6}{(Z_3 \cdot Z_5)} \qquad [7.3]$$

where C_z = 10.72 to 10.88, which accounts for the relationship of gear wheels with a fixed number of teeth.

For example, for $Z_3 = 46$; $Z_4 = 42$; $Z_5 = 26$; $Z_6 = 51$;

$$\text{we obtain: } P_u = \frac{10.72 \cdot 42 \cdot 51}{(46 \cdot 26)} = 19.199 \text{ threads/cm}$$

On Kovo pneumatic weaving machines, a positive fabric take-up mechanism with a periodic independent action (Fig. 7.2(c)) is used. The emery roller (1) is driven by gear wheels Z_3 and Z_2 from ratchet Z_1. Periodic movement is transferred to the ratchet by a pawl (2), set on the two-arm lever (3), controlled by means of two levers (4, 5), connecting rod (6) and cam (7). The fabric (8) is wrapped around the breast beam (9), emery roller (1) and a directing shaft (10), and moves on to the winding-up mechanism.

Figure 7.2(d) illustrates the Schönherr-type negative mechanism for periodic take-up of the fabric, where the action is dependent on fabric tension, such as is used on shuttle weaving machines for the development of woollen fabrics. The fabric (1) is taken up by the emery roller (2) using a directing platen (3) and arrives in the fabric box (4). The emery roller turns a ratchet (5) controlled by a moving pawl (6) mounted on the lever (7) carrying load G. Reverse rotation of the emery roller (2) due to the tension of the fabric K_T is prevented by a stop pawl (detent) (8). The rotation of the emery roller (2) is controlled by two moments: rotary moment M_m and resistive moment M_r, created by gravity G and the tension of the woven fabric K_T. From the equality of these moments, $M_m = M_r$, the traction effort of the regulator to fabric K_T is given by the equation:

$$K_T = \frac{G \cdot \ell_1 \cdot R}{(\ell_2 \cdot r)} \qquad [7.4]$$

In the absence of pressure from the reed (9) on the cloth fell (1), the moment of resistance $M_r = K_T * r$ should exceed the turning moment $M_m = G \cdot \ell_1 / \ell_2$. At beat-up, the reed (9) reduces the tension of fabric K_T and, hence, the magnitude of the moment M_r is less than M_m. The emery roller (2) takes up the fabric before equality $M_r = M_m$ is achieved. The thicker the weft yarns that are beaten up, the greater will be the angle of rotation of the emery roller. Thus, the fabric removal corresponds to the thickness of the weft, which is positioned in the fabric at approximately equal spacings b_i. The ratchet stroke is set by a limit stop (10) when the sley sword (11) is withdrawn from a cloth fell.

For shuttleless weaving machines, various companies have developed additional modules for programmed control of the amount of fabric take-up and, hence, the density of weft in the fabric. The take-up programme can be activated by the shedding device, or by a special device consisting of an electronic module and an electromechanical actuator.

7.2 Winding woven fabric on the cloth beam

After removal from the formation zone, the fabric is wound up on a cloth roller shaft (1) as shown in Fig. 7.3(a)–(e). The turning moment applied to the cloth roller can be created by means of friction forces T (Fig. 7.3(a)) of the fabric against the rough surface of emery roller (2), as found on shuttle weaving machines (Draper, AT). To increase the contact area of the fabric with the emery roller, all fabric winding mechanisms contain one or two directing platens (3). The force applied by the cloth roller (1) to the emery roller (2) is created by spring F. A large-diameter fabric roll cannot be wound by this type of direct action mechanism.

It is possible to wind a large-diameter fabric roll on an indirect-action mechanism (Fig. 7.3(b)) by using a separate frictional drive of the cloth roller (1) axis (1). Spring F here simplifies the drawing of the fabric onto the emery roller by means of two directing platens (3 and 4). On the Draper weaving machine, a roller mechanism is employed to wind the fabric (Fig. 7.3(c)). Of the two basic roller shafts (5 and 6) that are used, one (6) has a constant drive ω_1. As the diameter of the fabric roll increases, the axis of the cloth roller platen (1) moves on a side link lever (7), and the roll of fabric puts more pressure on the driving shaft, making the winding sufficiently dense . If the selvedges are thicker than the main fabric,, they may become distorted, leading to the defect known as 'wavy selvedge'.

On Sulzer weaving machines (STB) (Fig. 7.3(d)), the fabric roll is wound on the cloth roller (1) by means of an axial friction clutch consisting of a drive sprocket (4) onto the lateral surface of which friction coupling disks (5 and 6) are pressed by spring F. As the diameter D of wound fabric increases, the angular speed of its rotation will decrease according to the hyperbolic law. Therefore, the sliding of friction coupling disks (5 and 6) relative to the sprocket (4) will increase. Regulation of the force of spring F allows the driving moment M_m to be changed according to the required winding density: high with crossing (normal) selvedges or low with tucked-in selvedges.

Almost without exception, a winding diameter of up to 1 m can be accommodated on a cloth roller shaft (1) (Fig. 7.3(e)) outside the weaving machine, in the direction of the breast beam or the warp beam. On the Kovo pneumatic weaving machine, the fabric is passed by rollers (4 and 5) at ground level, under a guard below all the mechanisms of the weaving machine. The fabric roll is driven by a basic roller (7). This arrangement of the cloth roll (1) at the wide passage behind the back-rest allows a heavy roll of fabric to

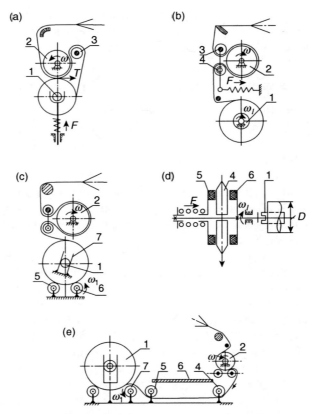

7.3 Mechanisms for winding of woven fabrics: (a) with drive from the emery roller (Draper); (b) with an axial frictional drive; (c) with a drive from platens (Draper); (d) with an axial drive (Sulzer); (e) arrangement of the roll at the back (Kovo).

be taken off mechanically. The fabric can, however, become soiled with oil or fluff.

7.3 Comparative analysis of methods of woven fabric take-up

The fabric designer should take into account the fact that fabric density depends on the weft (its 'thickness' or linear density (tex)) when comparing the actual structure of the woven fabric that is produced with a presumed structure. Programmed mechanisms regulating the cloth beam to give a constant or variable speed independent of the tension of the fabric can be used to create a variety of woven fabric structures. Analysis of the properties of the cloth beam regulator contributes to an understanding of the mechanisms for fabric take-up and for winding the woven fabric on the cloth roller. The cloth beam regulator of the Sulzer weaving machine has a versatile arrangement based on four interchangeable gear wheels to change fabric weft density.

7.4   Questions for self-assessment

1. What are the different arrangements in which the weft yarns can settle down in the structure of a woven fabric?
2. What are the distinctive features of fabrics having weft yarns distributed 'with a uniform distribution' and 'with a uniform beating-up'?
3. Under what conditions can fabrics with a uniform distribution of the weft yarns and with a uniform beating-up be produced?
4. Under what conditions is it impossible to achieve the formation of fabrics having weft with a uniform distribution and with a uniform beating-up?
5. What are the functions of the fabric take-up motion (cloth regulator)?
6. What are the factors on which the density of the weft yarns in a woven fabric depend?
7. What are the different methods of fabric take-up used on weaving machines?
8. What are the variants of fabric take-up mechanisms with independent discrete (intermittent) action?
9. What are the variants of fabric take-up mechanisms which have independent continuous operation?
10. What is the principle of operation of a fabric take-up mechanism which is dependent on the tension of the fabric woven?
11. What kinds of woven fabric structures can be obtained when using take-up mechanisms which are (i) dependent and (ii) independent of the fabric tension?
12. Identify the types of warp and weft yarns which are appropriate for use on weaving machines equipped with taking-up motions (cloth regulators) with various actions described in this chapter.
13. How it is possible to achieve a set density of weft in the fabric?
14. What are the features of the take-up roller (emery roller) used for withdrawing fabric from the fabric formation zone of a weaving machine?
15. What are the different methods of driving the emery roller?
16. What are the different methods of driving the cloth roller on a weaving machine?

7.5 References

1. Gordeev V.A. and Volkov P.V., 'Weaving', Leg. and Pitsh. Prom., Moscow, 1984 (in Russian).
2. Choogin V.V., Kahramanova L.F. and Nedovisiy M.N., 'Technology of Weaving Manufacture', State Technical University, Kherson, 2008 (in Russian).

8
Safety devices on weaving machines

DOI: 10.1533/9780857097859.135

Abstract: This chapter describes the safety (protective) devices provided on weaving machines: warp stop motions, weft detectors and safety devices against the breakage of warp threads.

Key words: safety devices, warp stop motions, weft controllers.

8.1 Introduction: safety devices on weaving machines

Safety mechanisms are provided on weaving machines primarily to prevent the occurrence of fabric defects (e.g. warp stop motions, wefts forks, weft feelers) and devices to detect faulty operation of moving parts. The devices for protection against mechanical damage of components are, as a rule, in the form of spring-loaded clamps which act to stop the weaving machine when a set level of force is exceeded.

To ensure the integrity of each warp thread in order to prevent defects in the fabric, such as floats or cracks, a warp stop motion is provided on all automatic weaving machines. These stop the weaving machine in the event of warp thread breakage. There are several types of warp stop motion: dropper and heald, mechanical or opto-electric (Ref. 1 and Ref. 2).

In shuttle weaving machines, the most extensively-used warp stop motion is the Taylor and Buckley system dropper with mechanical action, as found on Northrop weaving machines (Fig. 8.1(a)). In the event of a thread breakage, a dropper (1) falls into a gap between the teeth of two dropper rods, one of which is stationary (2), the other of which (2) is reciprocating. The movable lock-out rod (3) initiates a signal for the operation of the execution mechanism of a mechanical action (on shuttle weaving machines) or an electromechanical action (on shuttleless weaving machines) to stop the weaving machine.

On shuttleless weaving machines, the Beridot dropper warp stop motion with an electric activation is applied. A dropper (1) (Fig. 8.1(b)) falling as a result of a warp thread breakage closes an electrical circuit by means of an internal inclined feature that bridges the external electrode (2) and

136 Mechanisms of flat weaving technology

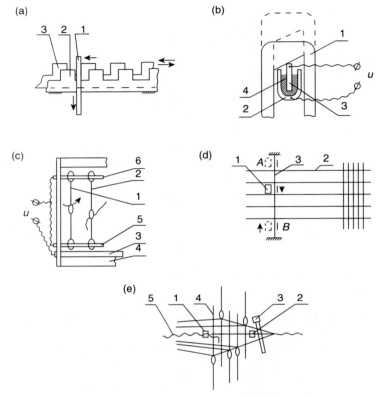

8.1 Warp stop motions. (a) Droppers with moving cogged bars (racks); (b) electric droppers (Beridot); (c) electric heald; (d) opto-electric system, with impulse counter; (e) opto-electric, radiant. Refer to text for detailed explanation of components.

the internal electrode (3), both of which are electrically isolated by insulator (4). A dropperless heald warp motion is presented in Fig. 8.1(c). During shedding, the tension of the warp threads keeps the healds (1) lifted in the bottom position of the harness. If a warp thread is broken, the heald (2) falls, making contact with the electrode (3) set on the bottom lath of the harness (4). This closes an electrical circuit which involves the metal heald carrier bars (5, 6) and the electrode (3) by means of metal healds (2). In the raised position of the heald shaft, warp tension will pull the healds down, which will cause them to make contact with the electrode (3). It is therefore necessary for the electrical circuit to be switched off during this period to prevent the weaving machine stopping. A disadvantage of this arrangement is that the weaving machine can only be stopped when a warp breakage occurs in the bottom position of the harness (4).

In the optical (contactless) method for detecting warp breakage, the warp threads (2) are monitored by means of an optical impulse counter (1) to check for any missing warp threads (Fig. 8.1(d)). The detector (1) makes back and forth movements on a guide (3) across the warp threads (2) and counts the number of threads. In the event of non-uniform distribution of threads, the counter is likely to malfunction, leading to the stoppage of the weaving machine. Also, the counter takes time to traverse the full width of warp of the weaving machine, and it is possible that, by the time a broken warp thread is detected, a defect (such as a crack or a float) may have formed.

The Hayashi (Güsken, Weko) dropperless contactless warp stop motion is of interest Fig. 8.1(e). Sources of light (1 and 2) and detectors containing phototransistors are located on a carrier along the two edges of the warp threads in one or more places (before and after heald device). One beam of light checks the cleanliness of the forward part of the shed between the reed (3) and the harness (4), and the second beam of light similarly checks the back part of the shed. In the event of a warp breakage, the broken thread (5) sags and crosses the light beam during the detection phase. This results in a signal to stop the weaving machine.

8.2   Weft controllers

An increase of labour productivity has been achieved in many respects by the introduction of weft forks, which are used to detect the presence of weft during the picking operation (Ref. 1 and Ref. 2). In shuttle weaving, the use of mechanical weft forks has been established in two variants: located at the edge of the shed, or inside the shed. Lateral weft forks (Diederichs) (placed on one side of the warp) can detect the presence of weft in the shed after insertion of the shuttle in the shed *after two picking operations*, and therefore the weaving defect called 'shed with missing weft' can occur. Centrally-located weft forks detect *every* weft inserted in the shed.

Figure 8.2 presents three illustrations of central weft forks for a shuttle weaving machine: (a) top view, (b) front view, (c) side view. The base (1) of the fork (2) (Fig. 8.2(a)) is firmly mounted on the sley (3) (Fig. 8.2(b)) by means of centre spindles (4). At the back of the sley (3) (Fig. 8.2(b)), the peg (5) of the base (1) is on a ramp *A* of a slider (6). Thus, the small tips (2) of the fork (Fig. 8.2(c)) are raised to allow for free flight of the shuttle.

When the sley moves to the fell of the fabric (7), the roller (8) (Fig. 8.2(a)) is moved up to the ramp (9) by a spring (10), which in turn moves the slider (6) by means of links to the right. Thus, the peg (5) (Fig. 8.2(b)), following the profile of the slider (6), turns on its centre spindles (4) (Fig. 8.2(c)) and lowers tips (2) into the shed. If the weft is taut, the tips 2 (Fig. 8.2(c)), pressing down on the weft (11), prevent rotation of the base (1), and the tooth *B*

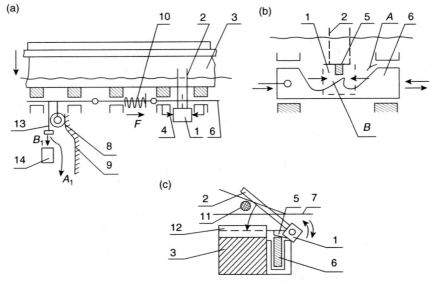

8.2 Central weft forks of shuttle weaving machine. (a) Top view; (b) front view; (c) side view.

(Fig. 8.2(b)) of the slider (6) passes under the peg (5). If weft is not present, or if there is a length of untensioned broken weft, the tips (2) (Fig. 8.2(c)) fall into the groove (12) of the sley (3), and the tooth *B* (Fig. 8.2(b)) rests against the peg (5). The movement of the roller (8) (Fig. 8.2(a)) to the right ceases and a pusher (13), with the further movement of the sley (3) to the cloth fell, presses on the link (14) of starting handles. The weaving machine stops.

For better control of the detection of unbroken weft, two forks on the width of the shed are used. The driving of the slider (6) can be carried out by a special cam.

Examination of the fabric for 'shed with missing wefts' on a shuttle weaving machine is made by turning the main shaft after stopping the weaving machine. A device for indicating the stoppage position, according to the number of weft yarns in the weave repeat is required on multi-shuttle weaving machines to preserve the accuracy of the weave. Identifying 'shed with missing wefts' demands a high level of skill from the weaver, since it is important to reduce the idle time of the weaving machine. Automatic devices can be incorporated which, following the detection of the condition 'shed with missing wefts' by a signal given from weft forks switch the dobby card to reverse its motion over between one and four individual turns of the main shaft as required, so as to find the shed with missing wefts and to stop the weaving machine with the sley in its back position. Removal of the broken end of the weft from the shed or its replacement by shuttle input in the shed is carried out manually.

A more rational solution is the application of a powerful brake to stop the weaving machine quickly, and a reverse motion sley mechanism to set the shed in its open position to enable the swift removal of the broken weft. In this way, the insertion of a faulty weft in the fabric is avoided, and also no time is lost searching for the fault. On shuttleless weaving machines, weft forks (Fig. 8.3) are established in several places: at the insertion side of the shed, in the middle the shed, or at the exit from the shed (Ref. 2). Weft forks differ structurally according to the principle of their action: mechanical, electromechanical, opto-electric, piezoelectric, triboelectric, high-frequency, etc. The effectiveness of the operation of the mechanism of weft forks depends on the complexity of the weft fork array: there are 30 on a Sulzer machine compared with three on the ATPR.

Figure 8.3(a) presents a diagram of the *lateral* weft forks on a Sulzer (STB) weaving machine. Following the picking of weft in the shed, a weft fork (1), held in place of a spring (2), checks for the presence of weft (3). The fork (1) is swung by a lever (4) connected to cams (5 and 6) by means of a link (7) and a lever (8), turning on the pivot O_1. The latter is controlled by a spring (9) and slide (10). When weft (3) is missing, fork (1), operated by the spring F, lowers a stop block A into the line of movement of stop block B. This leads to stop block B, the link (7), and hinge O_2 stopping. Tappets (5 and 6), pressing on rollers (11) move the lever (8) to the right on the pivot O_2. A slide (10) lifts the rod (15) by means of a ramp (12) and stops the weaving machine.

A very simple design of lateral weft fork is presented in Fig. 8.3(b). At the thread input into the sending rapier of the ATPR pneumatic-rapier weaving machine is a weft feeler (1) which is pressed by the weft thread (2). In the absence of weft, a spring (5) turns the feeler (1), closing electrical contacts (3 and 4). On rapier weaving machines, contactless action by a photo-electric 'feeler' is widely used (Fig. 8.3(c)). A beam of light (source (1)) is directed at a reflecting spot (2) on the head of the rapier (3). In the presence of weft (4), a lever (5) prevents the light from striking the spot (2). In the absence of weft, a spring (6) turns the lever (5) onto a stop block (7), exposing the spot (2) to the beam of light. The beam of light is then reflected onto the photodetector (1). The resulting electrical signal is used to stop the weaving machine.

On the pneumatic weaving machine (Elop, Elitex, Fig. 8.3(d)) a miniature electro-optical sensor is used in the air channel (1) which conveys the weft. When weft (2) is released into the shed through the channel, a beam of light (source (3)) falling onto a photo cell (4) is interrupted. If weft is absent, there will be no interruption of the beam of light. A steady beam of light leads to the stopping of the weaving machine.

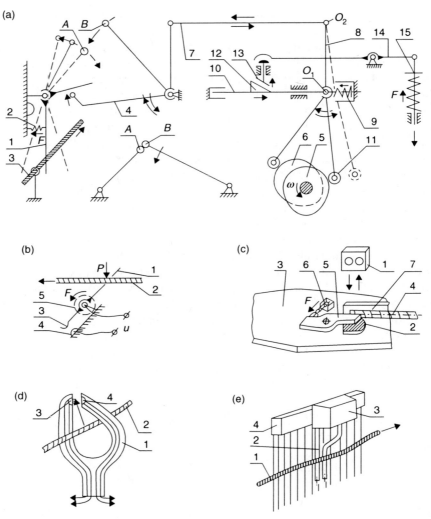

8.3 Weft forks on shuttleless weaving machines. (a) Lateral weft fork mechanical action (Sulzer: STB); (b) electromechanical action (ATPR); (c), (d) photo-electric action rapier (Rüti) and pneumatic (Elitex) weaving machine; (e) electric high-frequency hydraulic weaving machine, Nissan.

On water-jet weaving machines, damp weft (1) (Nissan, Enshu, Fig. 8.3(e)), bridging electrodes (2), closes an electrical circuit. The weaving machine stops in the absence of a thread (which creates an open circuit between contacts (2)). The base (3) of the contacts is fastened to the top part (4) of the sley. Triboelectric weft stops by Loepfe (Switzerland), use a ceramic sensor which produces a signal in response to the pressure of a the weft being

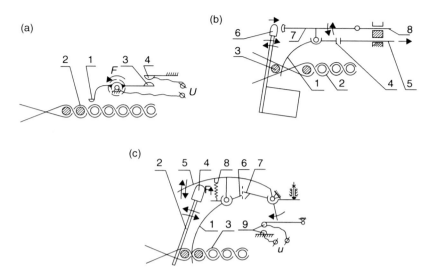

8.4 Weft feelers. (a) Cloth fell feeler; (b), (c) beaten-up weft feeler.

under friction as it moves. The signal is not affected by vibration or air dust content. On some weaving machines, weft forks are not fitted at the exit of the weft from the shed; therefore, short weft in the fabric can occur. On these weaving machines, weft feelers are positioned (Fig. 8.4) at the temple near the exit of the weft from the shed.

An elementary device of this type is found on the ATPR weaving machine (Fig. 8.4(a)). Here, a feeler (1) is pressed against the cloth fell (2) by a spring *F*. In the event of a missing weft in the fabric, the feeler (1) falls between the warp threads and its opposite shoulder closes electric contacts (3 and 4). This initiates a signal to stop the weaving machine. However, such weft feelers only work reliably after several weft yarns have not been incorporated into the structure of the fabric.

In the weaving of synthetic and metallic grids (Sulzer), a weft feeler (1) is used (Fig. 8.4(b)). It is lowered in front of the cloth fell (2) at the selvedges, which have a higher density of warp threads. Beating-up the weft (3) rotates the feeler (1) and its shoulder (4), which passes over a rod (5) pressing the top part of the sley (6) on an oscillating shoulder (7) of slider (8). If weft is absent, feeler (1) does not turn and the end of the shoulder (4) presses on the rod (5). This initiates a signal to stop the weaving machine. Feeler (1) is taken off by the shoulder (7) of the cloth fell (2), and returns to the beating-up line.

A more complex variant of weft feeler is used on pneumatic weaving machines. When the weft exits the shed, it has practically no tension (Fig. 8.4(c)). When the sley (2) moves forwards, a feeler (1) falls into the gap between the warp threads at the cloth fell (3). Following beating-up, the newly beaten-up weft has its end on the feeler (1) in the fabric. On the return movement of the sley (2), the top part (4), by means of a lever (5), lifts the feeler (1) from the fabric; the passage of the pawl (6) of the feeler (1) is controlled by the tooth of a lever (7). In the absence of a beaten-up weft, the feeler (1), under the influence of a spring (8), turns the pawl (6) to the right and, when the lever (5) rises on withdrawal of the sley, presses groove (flute) (7). This closes contacts (9) in an electrical circuit, which stops weaving machine.

8.3 Devices for prevention of warp thread breakage

In the event of a stoppage of the picking device in the shed, the sley will cause the breakage a group of warp threads at the cloth fell in its beating-up position over the length of the picking device (Ref. 1 and Ref. 2). Various devices have been used to prevent such yarn breakage and the resultant extended idle time of the weaving machine, and the inevitable flaws in the fabric: the collapsible reed and the key mechanism on shuttle weaving machines; devices to detect the presence of the picking device in the receiving box after its flight through the shed on shuttleless weaving machines.

On shuttle weaving machines, a collapsible reed is used (Bitschwiller). In the event of shuttle stoppage in the shed, the threads of the shed in the movement phase of the sley up to the cloth fell press the shuttle which, in turn, presses the reed, which initiates a signal to stop the weaving machine. To prevent the breakage of warp threads, various warp protectors have been devised (Löserson-Wilke, Lembcke and Döhmer, Reh, Diederichs); these dissipate the kinetic energy of the sley by striking shock-absorber links. The shuttle stops before approaching the cloth fell, thereby avoiding the rupture of warp threads.

Figure 8.5 shows a diagram of the warp protector widely used on automatic shuttle weaving machines. At normal picking through the shed, the shuttle (1), on entering the shuttle box, turns a valve (2) which, by means of a screw (3), operates on the arm (4), turns a key cylinder (5), and lifts the end of a stop rod rib (6). When the sley (7) moves to the cloth fell (8), the stop rod rib (6) passes over the picker receptor (9). In the event of stoppage of the shuttle (1) in the shed, the stop rod rib (6) is in its initial bottom position and, on the movement of the sley (7) will press against the picker receptor (9). The resulting impact against the picker receptor is absorbed by springs (10) by means of pressure from the core (11) and cushions (12). As a result of the movement of the cushion (12) by screw (13) by the pendant (14), the starting handle of the weaving machine will be returned to its 'off' position. This way, the sley is stopped *without* any warp thread breakage being caused by the shuttle.

Safety devices on weaving machines 143

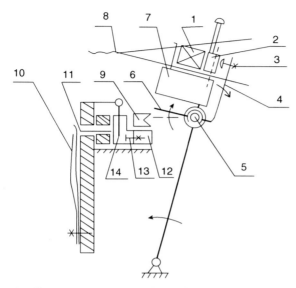

8.5 The warp protector of shuttle weaving machine (Reh, Diederichs, AT).

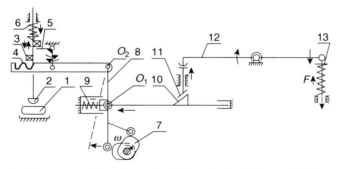

8.6 The controller of a weft picking device (Sulzer: STB).

Figure 8.6 presents a diagram of the mechanism which controls the arrival of the projectile in the receiving box of the Sulzer shuttleless weaving machine (STB). At normal picking through the shed, the projectile (1) does not allow the feeler (2) to fall. Thus, the ledge (3) stays above the line of action of the tooth of rack (4). In the event that the projectile (1) stops in the shed, the feeler (2) drops by means of a lever (5) and a spring (6). The ledge (3) slots into the groove of a stopping rack (4) and stops. Cams (7), continuing their rotation, turn the lever on the stopped hinge O_2, instead of hinge O_1 as during normal operation. Hinge O_1 moves to the left, overcoming the resistance of a spring (9). The cam (10) lifts a rod (11) by means of a lever (12), and a rod (13) initiates a signal to stop the weaving machine.

8.4 Comparative analysis of safety devices

The electrical Beridot dropper warp stop motion is relatively reliable and is widely used. The mechanism of lateral weft forks in the ATPR weaving machine have the simplest design and therefore have the greatest reliability (only three details).

8.5 Questions for self-assessment

1. What different purposes do safety devices serve on weaving machines?
2. What are the devices that have enabled the weaver's productivity to increase?
3. How does the warp stop motion influence the productivity of weaving machines?
4. When did mechanical weaving machines begin to be called 'automatic'?
5. How does the mechanical dropper type of warp stop motion differ from the electrical warp stop?
6. Why is the heald warp stop motion not widely adopted?
7. Why does the opto-electric warp stop motion have less reliability in the detection of warp thread breakage?
8. How does a lateral weft fork differ from a central weft fork on shuttle weaving machines? Which one of them can provide the higher quality of fabric?
9. What designs of weft fork are the most difficult to adjust properly on shuttleless weaving machines?
10. What type of weft fork do you recommend for use on shuttle and shuttleless weaving machines?
11. What is the purpose of the use of weft feelers on weaving machines of various designs?
12. Under what conditions will there be a breakage of groups of warp thread on a weaving machine?

8.6 References

1. Gordeev V.A. and Volkov P.V., 'Weaving', 'Leg. and Pitsh. Prom.', Moscow, 1984 (in Russian).
2. Choogin V.V., Kahramanova L.F. and Nedovisiy M.N., 'Technology of Weaving Manufacture', State Technical University, Kherson, 2008 (in Russian).

9
Weaving machine drives: mechanisms and types

DOI: 10.1533/9780857097859.145

Abstract: This chapter reviews the different types of weaving machine drives and stopping mechanisms of both shuttle and shuttleless weaving machines.

Key words: weaving machine drives, stopping mechanisms of weaving machines.

9.1 Introduction: the weaving machine drive

The drive of a weaving machine consists of an electric motor, the coupling to the main shaft of the weaving machine, the starting arrangement and the brake (Ref. 1 and Ref. 2). There are two drive configurations:

1. direct drive of the main shaft of the weaving machine by switching on the electric motor at start-up and switching it off to stop;
2. driving the main shaft of the weaving machine by means of a frictional clutch, with a constantly running electric motor.

A weaving machine drive should satisfy the following requirements:

- Due to the cyclical nature of the weaving process, the start and the completion of the operation of the mechanisms which carry out the main weaving functions should be strictly related to the corresponding angular positions of the main shaft.
- It should also be possible to start and stop the weaving machine at a low speed with the main shaft in various positions corresponding to the control mechanisms of the reed and harnesses, as required for the removal and repair of any broken weft and warp threads.

The need to use an electric motor capable of a high starting torque is an important requirement when direct drive is used on the main shaft of a weaving machine. In contrast, a drive with a frictional clutch enables the use of a constantly rotating large flywheel, facilitating the swift and smooth start-up of the weaving machine.

9.2 Mechanisms for driving the main shaft of a weaving machine

The principal types of main shaft drive mechanism for various weaving machines are presented in Fig. 9.1. The direct drive of the main shaft (1) (Fig. 9.1(a)) consists basically of the electric motor (2) and gear wheels Z_1 (the 'pinion') and Z_2 (the 'wheel'). The weaknesses of this system are the relative slow starting and stopping speeds.

An indirect drive with a friction disk (Fig. 9.1(b)) is popular on automatic shuttle weaving machines. The rotation of the electric motor (1) is transferred by gear wheel Z_1 carried on the motor shaft to the gear wheel Z_2 which carries the friction ring (2). The gear Z_2 *freely* rotates on a tube (3), which is splined to a shaft (6) with affords axial movement. When the actuating force P is applied by the starting device, the pressure disk (4) moves along the tube (3) and applies pressure on the rotating frictional ring (2), and displaces it with gear wheel Z_2 along tube (3), until contact is made with the thrust disk (5). This causes the closure of the friction clutch. The ring (2) imparts rotation to the thrust disk (5) and the main shaft (6). Springs (7) serve to separate the disks (4 and 5) following the withdrawal of force P.

On the Sulzer (also STB, ATPR) shuttleless weaving machine, the drive has a friction clutch; this consists of a cross (1) (Fig. 9.1(c) and (d)) with friction linings (2) and two half-clutches (3 and 4), driven by a belt drive from an electric motor. The cross (1) consists of four flexible steel plates, mounted on a splined hub (5), on the main shaft (6). Force P provided by the starting device displaces the freely rotating half-clutch (3) along the axis of the hub (5) until the friction liners on the ends of cross (1) are gripped firmly between it and the half-clutch (4). This results in the closure of the friction clutch, resulting in the rotation of the main shaft of the weaving machine.

In heavy shuttle weaving machines, a conical frictional clutch is used (Hartmann, Schönherr, Crompton, Rüti, (Fig. 9.1(e)), capable of transferring a high torque to the main shaft (1). The starting device provides a force P to shift the cone disk (2) which is integral with a pulley (3) carried on shaft (1). The pulley (3) is driven by a V-belt transmission, and when contact is established with the friction lining (4), the pulley (5) and the main shaft (1) rotate.

In Textima weaving machines, a drum-type friction clutch is used (Fig. 9.1(f)). The motor provides rotation to a pulley driven by a V-belt (1). The pulley (1) freely rotates on a splined tube (3) which is coupled to the main

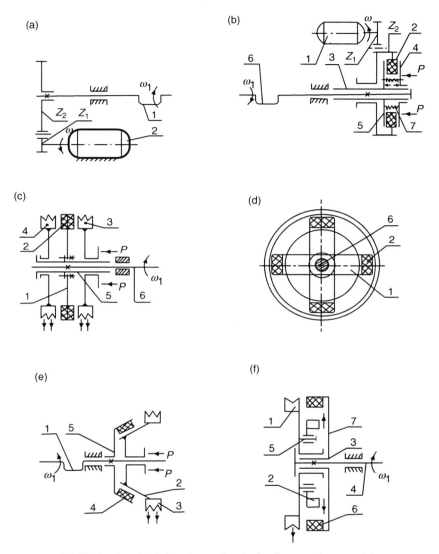

9.1 Methods of driving the main shaft of a weaving machine.
(a) Direct drive of the main shaft; (b) with a friction disk; (c), (d) Sulzer (STB); (e) Hartmann; (f) Textima. Refer to text for detailed explanation of components.

shaft (4). The starting device separates friction shoes (2) by means of fingers (5) and applies them against the internal surface of the circular friction lining (6) carried on the drum (7). The rotation of the drum is transferred to the main shaft (4) by means of a splined tube (3).

9.3 Weaving machine brakes

There are various types of brake to stop a weaving machine: electromagnetic, frictional shoe, disk or cone(Ref. 2). Older designs of brakes were based on a spring-operated single brake shoe which pressed onto a brake pulley externally (Atherthon Brothers, Hartmann). In order to increase the capacity of the brake, later designs employ two brake shoes, which are contained within the brake pulley.

In the same way as in a motor vehicle, two brake shoes (1) (Fig. 9.2(a)) with friction linings (2) are placed against a stationary circular support pin (3) and a cam (4). The actuating mechanism turns the cam (4) and applies the brake shoes (1) to the internal surface of the rotating pulley (5) which is mounted on the main shaft (6). This results in the braking of the shaft (6). At the start-up of the weaving machine, the brake is released

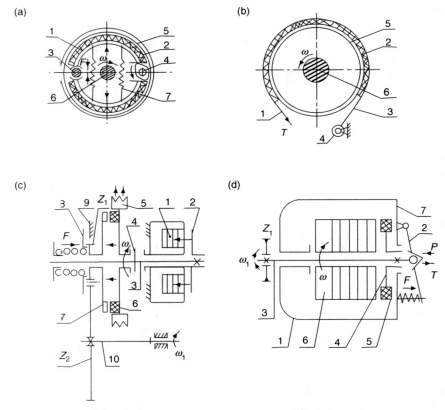

9.2 Weaving machine brakes. (a) Brake shoes; (b) band brake; (c), (d) combination mechanisms of drive and brakes (Novostav, Diehl).

by turning the cam in the opposite direction. Springs (7) cause the withdrawal of the brake shoes (1) from pulley (5). A brake with a floating axis (3) (i.e. one not fixed rigidly on the frame) such as the one provided on Elitex hydraulic weaving machines operates more accurately with even effort in each shoe.

The more powerful band brake (Fig. 9.2(b)) consists of a steel band (1) with a frictional lining (2) and a tensioning device operated by the starting mechanism. The end (3) of the band is fixed to a stationary anchor (4). At braking, the movable end of the band is stretched by links connected to the stopping device with a force T. The indicated direction of rotation of the pulley (5) attached to the main shaft (6) causes the energy of rotation to be dissipated by the frictional resistance applied by the band brake.

9.4 Combined start-up and braking mechanisms

Combined systems of weaving machine drive provide compactness of size and complete coordination of the start-up and braking phases (Ref. 2). In the Novostav weaving machine (Fig. 9.2(c)), the force of frictional closure of the half-clutch is created by an electromagnet (1). When activated, the stationary electromagnet (1) draws an armature in the form of a disk (2), which causes the displacement of the platen (3). The thrust washer (4) is displaced along the axis of the driving pulley (5). The friction liner (6) makes contact with the disk (7) of a half-clutch. The disk (7), overcoming the resistance of spring F, disengages the brake disk (8) from the stationary pad (9). Gear wheel Z_1, which is rigidly attached to the disk (7) provides rotation to the gear wheel Z_2 on the main shaft (10).

A friction clutch and a disk brake are integrated within the case (1) (Fig. 9.2(d)) of the Diehl electric motor system. By applying a force P on the lever (2) with the shaft (3), the mechanism is displaced along its axis. Thus, half-clutch (4) is pressed onto the freely rotating rotor (6) on the stationary shaft (3) by the friction ring (5). This results in frictional drive from the rotor (6) to the shaft (3) and, hence, gear wheel Z_1. When the lever (2) is moved in direction T by spring F, the shaft (3) is displaced to the right and the disk clutch (4) leaves the rotor (6) and the opposite face of the frictional ring presses against the stationary brake disk (7). The shaft (6) and gear wheels Z_1 stop.

9.5 Comparative analysis of different loom drives

- The weaving machine drive with a friction clutch in the Sulzer shuttleless weaving machine has the most rational construction.
- The band brake has the most powerful braking action.

- Combining the start-up and braking mechanisms of the weaving machine drive affords compactness and complete coordination of the start-up and braking phases, which ensures a high level of operational reliability.

9.6 Questions for self-assessment

1. What are the main types of mechanism of which weaving machine drives consist?
2. Which variant of the above would you recommend for use and why?
3. What are the requirements that a weaving machine drive should satisfy?
4. What is the basic difference between the two types of mechanism used in the drive of the main shaft of a weaving machine? Which type of drive is more appropriate for the weaving of heavy fabrics?
5. What are the basic differences between the braking mechanisms used on the main shaft of a weaving machine? Of these, which type is the most effective?
6. How is the problem of integration of mechanisms for driving and braking the main shaft of a weaving machine solved?

9.7 References

1. Gordeev V.A. and Volkov P.V., 'Weaving', 'Leg. and Pitsh. Prom.', Moscow, 1984 (in Russian).
2. Choogin V.V., Kahramanova L.F. and Nedovisiy M.N., 'Technology of Weaving Manufacture', State Technical University', Kherson, 2008 (in Russian).

10
Weaving machine parameters for specific woven fabric structures

DOI: 10.1533/9780857097859.151

Abstract: It is very important for the future specialist to know, all stages of analytical and experimental estimation of the optimal parameters of weaving machine settings with regard to specific woven fabric structures, the requirements of operational coordination of weaving machine mechanisms, the rational appraisal of weaving machine productivity in different units of measurement at different speeds of weaving machine operation according to requirements.

Key words: optimal parameters of weaving machine setting, productivity of weaving machines.

10.1 Introduction: the normalization process for weaving operations

The majority of woven fabrics are manufactured in significant volumes, typically in rolls of 10 000, 50 000 or 100 000 m. Manufacturers must maintain the technical parameters which determine the basic physical and geometrical properties of the specified fabric structure (thread parameters, woven structure, fabric elasticity, durability, etc.). This is a difficult challenge: many thousands of metres of one type of a woven fabric have to be manufactured on a considerable number of weaving machines (100, 1000 or more). However, weaving machines, even if they are all of one given type, may have individual peculiarities of operation, and so tend to form a fabric with significant differences in properties and quality (Ref. 1). These variations are likely to occur owing to the differences in:

- the adjustment of the weaving machines;
- the technical condition of the working mechanisms;
- the structure of the thread supply packages;
- the conditions of fabric formation.

The solution to this complex problem requires the normalization of the weaving process so as to provide identical conditions of formation of the given woven fabric (Ref. 2). *Normalization* of the weaving process involves:

- determining, analytically and experimentally, the optimum parameters of all elements of the elastic system of fabric formation (ESFF) and the rational parameters of adjustment of the basic working mechanisms;
- setting-up of drawing-in parameters on a given type of weaving machine which are optimum for a given type of fabric for the purpose of producing a good quality chosen structure of fabric and with the highest possible productivity.

For the optimal efficiency of the weaving process it is necessary to aim for three criteria:

- minimum level of thread breakages;
- maximum level of preservation of the initial physical properties of threads before being woven into the fabric;
- maximum level of productivity of the weaving machine.

The major factors defining the efficiency of the weaving process are:

- the parameters of shed geometry of the neutral line (NL), as defined by the relevant operating elements of the weaving machine;
- the level of tension of the elements of ESFF in the different phases of levelling, at open shed and during fabric formation;
- the degree of operational coordination of the basic working mechanisms of the weaving machine.

Templates and other such devices are used to achieve the setting up of rational drawing-in parameters. It is possible to recommend the following *technique of normalization* of the weaving process:

- **Stage 1** – As a rough guide, the calculation of the optimum drawing-in parameters by the use of formulae for a given type of weaving machine in order to produce the required fabric structure from a given yarn.
- **Stage 2** – Setting up the calculated drawing-in parameters of ESFF geometry relative to the NL on the weaving machine.
- **Stage 3** – Carrying out necessary adjustments, with the help of weaving trials, to minimize any non-uniformity of fabric structure during its formation.
- **Stage 4** – To carry out the evaluation of the magnitudes of optimum drawing-in parameters of the weaving machine with the use of yarn tension measuring equipment.

Weaving machine parameters for specific woven fabric structures 153

After the completion of the above four stages of weaving process normalization, it is necessary to carry out two finishing stages:

- coordination of the operations of the basic working mechanisms of the weaving machine;
- calculation of the productivity level of the weaving machine achieved.

At the beginning of the manufacture of a new woven fabric, it is essential not to be guided only by personal operating experience on weaving machines. The following specific sequence, established through engineering research for achieving optimum ESFF and machine–yarn–fabric path (MYFP) parameters of a weaving machine, is recommended.

10.2 Estimating drawing-in parameters for a weaving machine

The first stage of normalization of the process of weaving is to calculate the approximate values of all key parameters defining the conditions of formation and the quality of the fabric, as described below (Ref. 2).

1. The minimum drawing-in tension K_{min} of warp threads is that corresponding to the absence of any noticeable sagging of droppers and false stoppages of the weaving machine (and also the evidence of a clean shed). From experience (with weaving machines Sulzer, STB, ATPR, ELITEX, etc.), it can be established that, in the formation of an element of fabric, the tension of a single warp thread K_{bi} required at weft beat-up can be given by:

$$K_{min} = 0.4 \cdot K_{bi} + 10, \quad \text{cN/threads} \qquad [10.1]$$

This tension can then be set by carrying out adjustments to the warp regulator.

For example, for $K_{bi} = 50$ cN/thread, we obtain $K_{min} = 0.4*50 + 10 = 30$, cN/thread

2. At a given value of volumetric filling coefficient H_v (see Chapter 6, Equation [6.2]) of the specific structure of fabric on a weaving machine, a weft strip (cloth-fell) of length ℓ_u, mm will be formed on beating up:

$$\ell_u = C_b \cdot (12 \cdot H_v + 0.6) \qquad [10.2]$$

where C_b = the factor accounting for the type of fibre;

For cotton, $C_b = 0.8$; for wool, $C_b = 1.4$; for flax, $C_b = 0.6$; for silks, $C_b = 0.5$.

For example, for $H_v = 0.45$; $C_b = 0.8$. we obtain $l_u = 0.8 \, (12 \cdot 0.45 + 0.6) = 4.8$ mm.

3. For the purpose of optimization of conditions of formation of a stable fabric structure, it is necessary to define and apply a *levelling size* ℓ_b. For practical purposes, assuming the simple dependence of levelling size ℓ_b on coefficient H_v of volumetric filling of fabric on warp threads and weft, mm:

$$\ell_b = (0.674 \cdot H_v + 0.36) \cdot L_r . \qquad [10.3]$$

where L_r = the distance to the cloth-fell from the back position of reed, mm.

For example, for $H_v = 0.45$; $L_r = 69$ mm (Sulzer) we obtain
$\ell_b = (0.674 \cdot 0.45 + 0.36) \cdot 69 = 45.77$ mm

4. The difference of tension between the warp threads in the top and bottom branches of the shed is created by the *arrangement of height* of the back part of the shed by raising the dropper rail relative to the NL according to appropriate settings $+\Delta h_c$, 0, or $-\Delta h^1_c$ (see Chapter 3, Fig. 3.1), mm:

$$\pm \Delta h_c = C_3 \cdot (26.6 \cdot H_v - 4.0) \qquad [10.4]$$

where C_3 = the coefficient accounting for the nature of interlacing of threads in the fabric (for plain weave $C_3 = 0.7$; for twill, $C_3 = 1.0$; for satin/sateen, $C_3 = 0.2$; for derivatives of the three main weave interlacings, $C_3 = 0.4$).

For example, for $H_v = 0.45$; $C_3 = 0.7$.
we obtain $\Delta h_c = 0.7 \cdot (26.6 \cdot 0.45 - 4.0) = +5.58$ mm

5. The *position* of the dropper device by *depth* of the weaving machine defines the relationship of the tension of the spans of the warp threads in the front (ℓ_5) and back (ℓ_4) parts of the shed. The distance ℓ_{41} (the distance from the dropper device to the first heald shaft from the cloth-fell) can be given as, mm:

$$\ell_{41} = \ell_{51} \cdot A_d \cdot (2H_v + 1.7) \qquad [10.5]$$

where ℓ_{51} = the distance from the cloth-fell to the first heald shaft; A_d = a coefficient which depends on the specific features of construction of the weaving machine.

For example, for weaving machine Sulzer (STB) $A_d = 1.0$; for rapier (R-190), $A_d = 1.3$; for pneumatic (P-155), $A_d = 1.4$; for pneumatic-rapier, (ATPR) $A_d = 1.6$.

For example, for $\ell_{51} = 137$ mm; $H_v = 0.45$; $A_d = 1.0$, we obtain

$$\ell_{41} = 137 \cdot 1.0 \cdot (2 \cdot 0.45 + 1.7) = 356.2 \text{ mm}.$$

6. The opening angle of the front part of the shed (2α) determines the length of the shed and, hence, the extension of warp threads at shedding and the likelihood of their breakage. It is possible to use the simple empirical formula:

$$2\alpha = K_h \cdot \alpha_{min}. \qquad [10.6]$$

where K_h = the coefficient depending on the type of fibre of the warp threads (for cotton, $K_h = 1.15$; for silk, $K_h = 1.0$; for wool, $K_h = 1.25$; for flax, $K_h = 1.35$); α_{min} is the minimum acceptable angle of the shed on a weaving machine of a given construction (ATPR, $\alpha_{min} = 28°$; Sulzer, $\alpha_{min} = 16°$; Elitex, $\alpha_{min} = 20°$). For example, for $K_h = 1.15$; $\alpha_{min} = 16°$ we obtain $2\alpha = 1.15 \cdot 16 = 18.4°$.

10.3 Setting up parameters for the machine–yarn–fabric path (MYFP) on weaving machines

The second stage of normalization involves the setting up of calculated parameters of the *geometry of* the MYFP regarding the NL of the shed on a given weaving machine by means of the device shown in Fig. 10.1 (Ref. 2). This apparatus facilitates the setting up of the appropriate parameters of the yarn and fabric path on the weaving machine. It consists of the base-ruler (1) and its strengthening support (2). Movable supports (3, 4 and 5) are carried on slides along the base (1). The supports (4 and 5) can be adjusted for height in slides (6 and 7). Millimetric divisions are marked along the length of the base (1) and on the slides (6 and 7). The length of all supports is identical. In use, the support (2) is secured on the cloth-fell at point A. Coloured thread (8) is passed (drawing-in) through an eye heald (9) in the level phase and fixed in a tensioned state on the end of the support (3). The base (1) will occupy a position parallel to the NL of the shed NL at the arrangement of the thread (8) *in the middle of* a heald eye (9). With the support (2) and the base (1) in this position, it is possible by means of supports (4 and 5) to define the array parameters of the dropper device (point C) and the lines of contact of threads of the back-rest (point D) on the depth and height of the shed relative to NL: $\pm h_d$; $\pm h_c^1$; ℓ_c; ℓ_D.

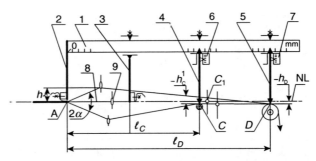

10.1 Apparatus for setting up the parameters of the MYFP on a weaving machine. Refer to text for detailed explanation of components.

10.4 Verification of parameters for the elastic system of fabric formation (ESFF) and machine–yarn–fabric path (MYFP) on weaving machines

At the third stage of normalization of the MFYP and ESFF, it is recommended that the objective experimental verification of the calculated optimum drawing-in parameters of weaving is carried out. This involves the checking for non-uniformity of warp threads interlacing in the fabric under the dynamic working conditions of the weaving machine (Ref. 2 and Ref. 3). Any error in preparation of the threads for weaving and the mechanical operation of the weaving machine are reflected in the degree of uniformity of loading distribution of the separate warp threads. There is no equipment available for the simultaneous adjustment of the tension and stretch deformation (extension) of *each* single warp thread on the weaving machine (to achieve this, it would be necessary to use over one thousand tiny tension gauges!). Therefore, in practice, it is recommended that simple and *objective rapid quality monitoring* of the non-uniformity of interlacing of warp threads in the fabric is used.

This procedure is as follows:

- The weaving machine is stopped. On the warp beam, a line of a contrasting colour is drawn across the warp threads parallel to the beam axis in a position which is about 300° in front the position where the warp threads come off the warp beam.
- After restarting the weaving machine for normal weaving, the marks on the warp threads gradually move over the back-rest, then past the droppers, heald shafts, then into the cloth-fell and become woven into the fabric.
- The distribution of the marks (Fig. 10.2), the amplitude B_i and the form of distribution of marks on various points of the fabric (in the zone of

Weaving machine parameters for specific woven fabric structures

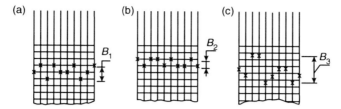

10.2 Non-uniform crimp of warp threads in the woven fabric (a), (b), (c) With average, narrow and wide spacing of marks in the fabric.

the temples, in the middle of a fabric, and at the selvedges) defines the deviation from the norm B_N.
- The analysis of the reasons for the distribution of the marks (defect of winding and warping, non-uniform impregnation of warp yarns in the sizing machine, poor-quality winding on the warp beam, entanglement or sticking of threads in the shed, fallen droppers, weakened harness, jamming of temple rings, etc.) can be now made.
- Measures for the avoidance of the identified defects of procedure or the equipment can then be taken.

A skilled production operative can master the above objective method sufficiently quickly.

10.5 Evaluating warp thread tension by oscillogram analysis

At the fourth stage of normalization of woven machine and working conditions, it is advantageous to use appropriate means for evaluating the weaving process; such as electronic equipment for recording warp thread tension. Use of the *tension measuring* method by graphical representation of processing changes of the tension of warp threads in the ESFF is a relatively laborious and expensive process (Ref. 3 and Ref. 4). Highly-skilled, experienced personnel are needed to evaluate the optimality of drawing-in parameters by means of electronic equipment and complex tension gauges on groups of warp threads and the weft, the measurement of beat-up forces, cloth-fell displacement (the amount of movement of the cloth-fell by the reed during beat-up), etc. However, by examining the geometrical form of an oscillogram of variation in the tension of warp threads during formation of *all repeat interlacings* of threads in a fabric, it is possible to gain detailed information on the working conditions of threads at any value of the angle of rotation of the main shaft of the weaving machine.

It is necessary to pay close attention to the distinguishing forms of the shape of variations of warp thread tension in different phases (opening of

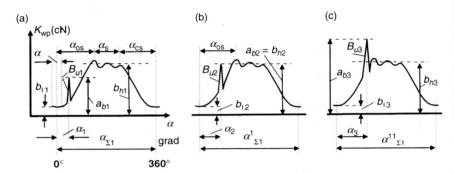

10.3 Idealized models of tension variation of warp threads for one rotation of the main shaft of a weaving machine $\alpha_{\Sigma 1} = \alpha_{1\Sigma 1} = \alpha_{11\Sigma 1} = 360°$. (a) Beat-up in the starting phase of shed opening; (b) beat-up in the mid-phase of shed opening; (c) beat-up towards the end of shed opening.

the shed, weft beating-up, opening the shed and the picking of the weft, closing of the shed and the level phase of the warp threads). The ability to capture and interpret such *detailed* information enhances the tension measuring method identified here by enabling the *highest level of control* of the conditions of development of a fabric of a specific structure.

In practice, tension oscillograms are recorded on tape or paper of limited width, owing to which, for each structure of fabric, it is necessary to use a specific scale (cN/mm) for the oscillogram showing the variation of the magnitude of tension of the warp threads K_{wp}. Therefore, for the visual comparison of various tension oscillograms, it is important to take this specific scale into account. The following method is found to be useful for the analysis of typical oscillograms of tension variation of warp threads on the formation of woven fabrics of various densities (Figs 10.3–10.6).

Figure 10.3(a), (b) and (c) show a simplified model of the variation of the tension of warp threads for one rotation of the main shaft of the weaving machine $\alpha_{\Sigma 1} = \alpha^1_{\Sigma 1} = \alpha^{11}_{\Sigma 1} = 360°$ at three variants of the shape of the beat-up peak a_{bi}. The minimum tension of warp threads occurs in the levelling phase. In the given example, $b_{f1} = b_{f2} = b_{f3}$ (cN). The opening phase of the shed lasts an angle of rotation α_{os} of the main shaft of the weaving machine. It also shows the dwelling phase of the heald shaft and warp threads in the fully opened shed (α_s degrees) and the shed closing phase ($\alpha_{cs} = \alpha_{os}$ degrees).

In the opening of the shed, thread tension essentially increases from $b_{f1} = b_{f2} = b_{f3}$ to $b_{h1} = b_{h2} = b_{h3}$ (cN). Distortion of the form of the curve (tension oscillogram) at shed closing occurs when the weft is beaten up by the reed into the fell of the fabric. There is the 'beat-up peak' B_{ui}. In the first possible variant (Fig. 10.3(a)), the beating-up of the weft is carried out as the shed begins to open at α_1 (degrees). The beat-up peak B_{u1} has the magnitude a_{b1} (cN), which

Weaving machine parameters for specific woven fabric structures 159

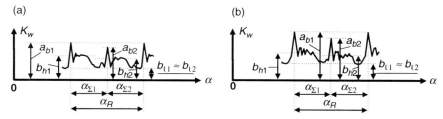

10.4 Tension oscillograms of warp threads on the formation of repeat interlacing of a dense fabric of plain weave of average strain ($H_v \approx 0.45 \div 0.55$) under unequal tension in the shed branches. (a) Stable process of forming woven fabric and (b) abnormal variant of ESFF parameters.

is less than the tension of warp threads at open shed, b_{h1}. This variant of warp thread tension oscillogram occurs when developing very 'easy' fabrics of low density with a small volume filling factor in the warp and weft ($H_v \approx 0.2$).

At the second possible variant (Fig. 10.3(b)), the beating up of weft B_{u2} is carried out at shed opening on a position $\alpha_2 \approx 0.5\alpha_{os}$ (degrees). The tension of warp threads at beating-up is approximately equal to the tension of threads in the fully opened shed $a_{b2} \approx b_{h2}$ (cN). This variant of warp thread tension oscillogram happens when developing 'average' fabrics with a volume filling factor in the warp and the weft ($H_v \approx 0.3$–0.4).

At the third possible variant (Fig. 10.3(c)), the beating-up of weft B_{u3} is carried out at the end of shed opening phase at a position α_3 before α_{os} (degrees). The tension of threads at beating-up is higher than the tension of threads in the completely opened shed $a_{b3} > b_{h3}$ (cN). Such a variant of warp threads tension oscillogram occurs when developing dense fabrics with a higher value of factor of volume filling in the warp and weft ($H_v \approx 0.5$–0.6).

In the subsequent figures (Figs 10.4–10.6), examples of **typical** tension oscillograms in weaving cotton warp threads are presented reflecting the formation of fabrics of various density and volume filling values H_v. The comparative analysis of these tension oscillograms gives the opportunity to define a measure (degree) of rationality of conditions of formation of a fabric, and also to identify inadequacies of adjustment of the weaving machine for the development of the fabric of set volume filling H_v.

In Fig. 10.4, two warp tension oscillograms are represented reflecting the repeat interlacing of a dense fabric of plain weave of average intensity with a shedding cycle 1/1 and with a factor of volume filling from cotton threads $H_v \approx 0.45$–0.55. In analyzing the tension oscillogram in Fig. 10.4(a), it is possible to state the following:

- The cycle of development of repeat interlacing of threads is equal to two rotations of the main shaft $\alpha_R = (\alpha_{\Sigma 1} + \alpha_{\Sigma 2}) = 720°$.

- In the level phases of the first ($\alpha_{\Sigma 1}$) and second ($\alpha_{\Sigma 2}$) rotations of the main shaft, the warp threads have similar magnitudes of tension $b_{\ell 1} \approx b_{\ell 2}$.
- During the open shed phase on the second ($\alpha_{\Sigma 2}$) rotation of the main shaft, the tension of the warp threads is less than on the first $b_{h2} < b_{h1}$ (about $\approx 10\%$).
- At the beating-up of the second weft ($\alpha_{\Sigma 2}$), the tension of warp threads is also less than on the beating up of the first $a_{b2} < a_{b1}$ (about $\approx 13\%$).
- Comparing the magnitudes of tension of the warp threads at beating-up and at open shed, it is possible to ascertain their essential inequality to be $\approx 37\%$ ($a_{b1} > b_{h1}$) and $\approx 35\%$ ($a_{b2} > b_{h2}$).

From the analysis of tension oscillogram (Fig. 10.4(a)) it is possible to draw the following conclusions:

- The unequal sizes of tension of warp threads in the open shed $b_{h2} < b_{h1}$ specifies an unequal tension of the top and bottom *branches* of the shed. In practice, this state can be achieved by installation of the dropper device and back-rest above or below the NL of the shed.
- When the tension of the branches of the shed is unequal, every second weft is beaten into the fell of the fabric with reduced beat-up force ($a_{b2} < a_{b1}$).
- Every first weft of each repeat of thread interlacing is beaten-up at the raised tension of warp threads that leads to a denser arrangement at the fell of fabric.
- A generally steady form of tension oscillogram indicates a reasonably high degree of stability of the arrangement of the branches of the shed on picking the weft and, on the whole, a stable process of formation of *repeat* thread interlacing in plain-weave fabric accompanied by a reduced beat-up force.

On the other weaving machine (Fig. 10.4(b)) developing the same plain-weave fabric, the warp tension oscillogram shown in the figure is located further away from the zero line α in comparison with the tension oscillogram in Fig. 10.4(a). It specifies a higher level of tension of warp threads in all phases: levelling, open shed and beating-up. In particular, in comparison with the tension oscillogram in Fig. 10.4, the tension of warp threads in the level phase has increased, by approximately the same amount (by $\approx 60\%$); in the phases of the open shed, by $\approx 20\%$ and by $\approx 22\%$; and in the beating-up phase, by $\approx 25\%$ and $\approx 28\%$.

The analysis of the tension oscillogram (Fig. 10.4(b)) reveals the presence of sharp fluctuations of warp tension with a higher amplitude during open shed periods. Such fluctuations of thread tension are the consequence of improper adjustment of the weaving machine. This could be because of the reduced spring tension of the movable system (mechanism) of the back-rest and improper adjustment of the warp beam regulator. The high amplitude of fluctuation in the tension of the shed branches can complicate weft picking and could lead to the breakage of warp threads. If this is the case, the weaving machine service engineer could be asked to lower the primary setting tension to the level b_{ti} in the tension oscillogram, as in Fig. 10.4(a).

Figure 10.5 presents four variants of tension oscillograms of warp threads, as found in weaving very dense plain-weave fabrics with a volume filling $H_v = 0.70 \div 0.90$ (canvas, heavy technical, etc.). Here, it is necessary to pay special attention to an important matter: all tension oscillograms of warp threads have a beating-up peak of a very high value. It specifies the formation of a fabric with a highly-increased cloth-fell length (up to 10 mm).

Figure 10.5(a), shows a tension oscillogram of warp threads in the level phase, which very closely approaches the zero line α at each rotation of the main shaft of the weaving machine. It indicates the likelihood of false stoppage of the weaving machine by operation of the dropper mechanism owing to (considerable) unacceptable lowering of the droppers, leading to frequent stoppages due to the closure of the electrical circuits of the warp

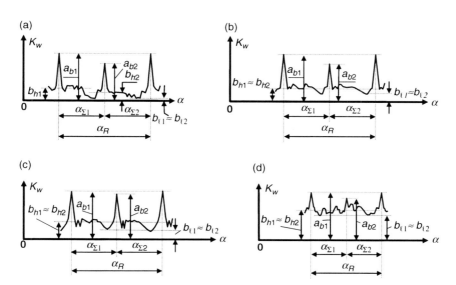

10.5 Tension oscillograms of warp threads on the formation of repeat interlacings of a compact plain-weave fabric of higher strain ($H_v \approx 0.70 \div 0.90$). (a) Lower drawing-in tension and unequal tension of shed branches; (b) more stable process of forming woven fabric where the tension of the shed branches is unequal; (c) optimum condition of fabric forming where the tension of the shed branches is equal; (d) excessively minimal tension of warp threads.

thread breakages detection system. In the open shed phases, at beating-up, a clear inequality of warp tension is observed: $b_{h2} < b_{h1}$ to $\approx 40\%$ and $a_{b2} < a_{b1}$ to 21%. It indicates the development of the fabric with a shed of unbalanced (unequal) tensions.

As shown by Fig. 10.5(b), in the open shed phase of the first and the second weft picking, an equality of tension of threads $b_{h1} \approx b_{h2}$ is reached by increasing the minimum necessary tension in the level phase $b_{\ell i}$ in comparison with $b_{\ell i}$ in Fig. 10.5(a). In the beating-up phase, in the first weft (over $\alpha_{\Sigma 1}$) the amount of tension of threads a_{b1} has remained at the previous level (as in Fig. 10.5(a)). On the second rotation $\alpha_{\Sigma 2}$ of the main shaft of the weaving machine, the value of the beat-up peak at the second weft insertion has decreased, $a_{b2} < a_{b1}$ by $\approx 15\%$ (instead of $\approx 21\%$ as in Fig. 10.5(a)).

By comparative analysis of these two tension oscillograms, it is possible to draw the following conclusion: the service engineer has stabilized the process of formation of the heavy plain weave by a significant increase in the minimum tension of the warp threads. But there could be negative consequences due to the increased degree of *unbalance* in the tension of the shed branches. A higher level of stability in the formation process of the same structure of fabric can be achieved by transition from an unbalanced shed tension to one of balanced tensions (Fig. 10.5(c)). In the figure, the uniformity of tension of warp threads in the levelling and open shed phases is clearly visible at each rotation of the main shaft of the weaving machine: $b_{\ell 1} \approx b_{\ell 2}$ and $b_{h1} \approx b_{h2}$. The variation of the beating-up force, a_{b1} and a_{b2} has decreased to 5%. Under these conditions, the fabric formed has a uniform structure and a flat (even) surface.

The conclusions drawn from the comparison of tension oscillograms (a), (b) and (c) visually specify the optimum variant (c) of adjustments to the ESFF and MYFP, providing stable conditions for forming a heavy plain-weave fabric by the creation of equal tension in the top and bottom *branches* of the shed at an optimum (effective) level of the minimum primary setting tension of the ESFF. As an example of the improper adjustment and state of the basic mechanisms of a weaving machine, the tension oscillogram of warp threads is presented on development of the same structure of fabric, as in Fig. 10.5(d). From the shape of the warp threads tension oscillogram, it is possible to ascertain the presence of:

- incorrect adjustment of the warp beam regulator;
- a worn out of the tappet shedding device and take-up motion (cloth regulator), as indicated by mechanical play between components, and a reduced range of movement of working elements (heald shaft, emery roller, etc.);
- more frequent breakage of warp threads and, hence, increased idle time owing to excessively high warp tension fluctuations.

Weaving machine parameters for specific woven fabric structures

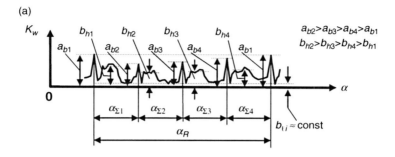

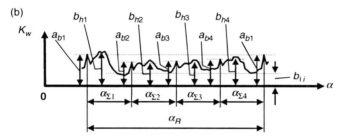

10.6 Extreme unbalanced b_{li} tension oscillograms of warp threads on the formation of fabric of average intensity: type 'Twill 1/3' with factor of volume filling $H_v \approx 0.50$. (a) Low minimum tension of warp threads; (b) excessively minimal tension of warp threads.

It is obvious that the service engineer has increased the minimal tension by 2.5 times (in comparison with tension oscillogram in Fig. 10.5(c)), with the intension of eliminating the above deficiencies. However, this action has led to an excessive level of warp thread tension, with the attendant consequences (breakage of warp threads would have increased due to overstraining, and the fabric produced would be of reduced strength).

Having demonstrated how to analyze the geometry of a typical warp thread tension oscillogram in the development of plain-weave fabrics so as to obtain optimum quality and production, two extreme examples of a warp tension oscillogram are presented. These reflect less well-connected, less dense interlacing of a fabric with long floats of type 'Twill 1/3' with a factor of volume filling $H_v \approx 0.50$ (Fig. 10.6). The full cycle of development of interlacing of twill 1/3 occurs over four rotations of the main shaft of the weaving machine: $\alpha_R = \alpha_{\Sigma 1} + \alpha_{\Sigma 2} + \alpha_{\Sigma 3} + \alpha_{\Sigma 4}$.

As shown by Fig. 10.6(a), at the first rotation of the main shaft of the weaving machine $\alpha_{\Sigma 1}$ it is found that:

- shed branches in the picking phase of weft threads have a tension with maximum level b_{h1};

- the beating-up of the first weft of repeat interlacing also occurs at the highest level a_{b1} of warp thread tension.

The second weft is inserted in the shed and beaten in to the fell at the lowest tension of warp threads (a_{b2} and b_{h2}). The third and the fourth wefts are picked into the shed and beaten-up to the fell of the fabric with a higher level of warp tension. With reference to the given tension oscillogram, the difference in tension of warp threads in the open shed phase (during weft insertion), starting from tension b_{hi} in the beating-up phase a_{bi} consistently decreases: $b_{h1} < a_{b1}$ by 44%, $b_{h2} < a_{b1}$ by 24%, $b_{h3} < a_{b1}$ by 22%, $b_{h4} < a_{b1}$ by 11%.

The warp thread tension in the open shed position b_{hi} gradually increases in comparison with the minimum b_{l2} at the insertion of the second weft in the shed: $b_{h3} > b_{h2}$ by 20%, $b_{h4} > b_{h2}$ by 33%, $b_{h1} < b_{h2}$ by 83%. The range of change in tension is equal to 83%, 12% = 71%. Thus, the tendency of a gradual increase in the warp thread tension in weft insertion phases $b_{h2} < b_{h3} < b_{h4} < b_{h1}$ and beating-up phases $a_{b2} < a_{b3} < a_{b4} < a_{b1}$ in the process of formation of repeat of a fabric interlacing is observed at durations of $\alpha_{\Sigma2}$, $\alpha_{\Sigma3}$, $\alpha_{\Sigma4}$ and $\alpha_{\Sigma1}$.

It is interesting to note the stability of the magnitude of the minimum necessary tension of warp threads $b_{l1} \approx b_{l2} \approx b_{l3} \approx b_{l4}$. However, the level of tension of warp threads b_{li} in the level phase appears to be very low. Such reduction in the warp thread tension in a cycle of shedding can lead to warp threads sticking, resulting in the formation of an uneven shed and to the increase of warp thread breakage in the shed on weft insertion. The conclusion to be drawn from the analysis of the given tension oscillograms is that the geometrical form of tension variation of warp threads on repeat interlacing may indicate irrationally low levels of the minimum tension of warp threads.

Another extreme case of minimal warp thread tension in weaving a twill 1/3 is presented in Fig. 10.6(b). From the geometry of tension oscillogram, it is seen that the minimum level of tension has essentially increased (by about 3.5 times) in comparison with the tension oscillogram given in Fig. 10.6(a). Owing to such high increases in of warp thread tension in the levelling phase (b_{li}), thread tension in the open shed phase, b_{hi}, has also increased. It is necessary to note that there will be an unavoidable increase of warp thread tension b_{h1} at the picking of the first weft on repeat fabrics by 61 per cent in comparison with b_{h1} in Fig. 10.6(a). Also, size b_{h1} has exceeded even size a_{b1} by some 10 per cent (Fig. 10.6(b)).

In comparison with the minimum level of warp thread tension in the open phase of shed b_{h2} (on the second rotation of the main shaft) the following inequalities are observed: $b_{h3} > b_{h2}$ by 14%, $b_{h4} > b_{h2}$ by 23%, $b_{h1} < b_{h2}$ by 46%. The range of increase in tension is equal to 46% – 14% = 32%. As a whole, there is an increase of warp thread tension in levelling phases b_{li}, and the open shed b_{hi} has not led to an increase a_{bi}. On all rotations of the main shaft of the weaving machine, values $\alpha_{\Sigma 1}$, $\alpha_{\Sigma 2}$, $\alpha_{\Sigma 3}$ and $\alpha_{\Sigma 4}$ of beating peaks a_{bi} remained at their former level. Thus, beat-up tension has not changed, despite a large increase in the minimum tension of the warp threads. As a result of the analysis of the tension oscillograms in Fig. 10.6(a) and (b), it is possible to recommend the following: to optimize production of the given twill 1/3, it is necessary to lower the level of the minimum tension of warp threads $b_{\ell i}$ in Fig. 10.6(b) by 30 per cent.

In practice, a considerable number of different types of weaving machines with mechanisms of various constructions are used in the production of many fabric structures. Therefore, the examples given above of geometrical forms of the variation of tension of warp threads on weaving machines during the production of fabrics of various densities (volume filling by warp and weft threads) illustrate the complexity of the process of analyzing tension oscillograms and formulating recommendations to improve fabric production and quality.

However, it is possible to offer some *general recommendations* regarding the normalization of production processes for any fabric:

1. During the weft insertion phase (i.e. open shed), it is preferable to have a warp tension with the least possible range of fluctuation.
2. In the level phase of the shed, a minimum level of warp tension should be maintained; tension should be sufficient to prevent droppers from sagging, but also enable reliable division of warp threads at shed crossings, which is important for the avoidance of snagged threads.
3. Due to the design of a fabric and its intended use, it is possible to identify two variants of tension of the shed branches:
   - unequal tension of the two shed branches is useful as it permits a reduction in the necessary beat-up force;
   - equally-tensioned (balanced) shed branches stabilize the array of all weft in the structure of fabric.

The ability to analyze a warp tension oscillogram make efficient recommendations, and gaining confidence in those skills, can only be developed by broad practical experience at a weaving firm.

10.6 Coordination of weaving cycles

It is convenient to indicate the coordination of the operational phases of all weaving machine mechanisms according to a joint cycle example for 360° or 720° degrees of rotation of the main shaft of the weaving machine. Such a cyclogram can be obtained in a circular or linear form. For shuttleless weaving machines, a linear cyclogram is preferable; this contains the basic working mechanisms (or operations), and the broken lines define the operational course. It is *convenient* for technologists to take the traditionally accepted extreme front position of the reed in the fabric formation phase as being the initial position (zero).

10.7 Factors affecting the productivity of weaving machines

At the final stage of normalizing conditions for fabric formation, it is necessary to define and make a rational estimate of the possible performance level of a weaving machine at various speeds of rotation of the main shaft, in suitable units of measurement (Ref. 2). As the theoretical productivity of weaving machine P_T, we estimate the quantity of fabric produced by the weaving machine, without making allowances for idle time, during the course of one hour, in m/h:

$$P_m = \frac{60 \cdot P_T \cdot n \cdot i}{100 \cdot P_u} \quad [10.7]$$

where n = the frequency of rotation of the main shaft of the weaving machine, rpm;

i = the number of picks inserted in one cycle of fabric formation (for one rotation of the main shaft of the weaving machine);

P_u = weft density in the fabric, threads/cm.

For example, for $n = 280$ rpm; $i = 1$; $P_u = 25.0$ threads/cm

$$\text{we obtain } P_m = \frac{60 \cdot 280 \cdot 1}{100 \cdot 25.0} = 6.72 \text{ m/h.}$$

Actual productivity P_{fL} of a weaving machine is defined by the quantity of the fabric made by the weaving machine in a unit of time, taking into account the idle time, m/h:

$$P_{fL} = P_m \cdot \eta_L \quad [10.8]$$

where η_L = fraction of useful operation, over the operating time of the weaving machine.

For example, for $P_T = 6.72$ m/h; $\eta_L = 0.88$, we obtain

$$P_{fL} = 6.72 \cdot 0.88 = 5.91 \text{ m/h.}$$

Weaving machine parameters for specific woven fabric structures 167

For comparison of productivity of weaving machines developing fabrics with various densities of weft, productivity is expressed in thousands of picks/h:

$$P'_{fL} = \frac{60 \cdot n \cdot i \cdot \eta_L}{1000} \qquad [10.9]$$

For example, using the above figures we obtain,

$$P_T = (60*280*1*0.88/1000) = 14.784 \text{ thousand weft/h.}$$

To compare productivity of weaving machines with various widths of warp, developing the same fabric, it is convenient to express productivity in square metres, m²/h:

$$P''_{fL} = \frac{60 \cdot n \cdot i \cdot B \cdot \eta_L}{100 \cdot P_u} \qquad [10.10]$$

where B = the width of a fabric, m.
For example, for n = 280 rpm; i = 1; B = 1,59 m; η_L = 0.88 ; P_u = 25.0 threads/cm,

$$\text{we obtain } P''_{fL} = \frac{60 \cdot 280 \cdot 1 \cdot 1.59 \cdot 0{,}88}{100 \cdot 25.0} = 9.4 \text{ m}^2/\text{h}$$

At a factory, the productivity of a weaving machine in meters of weft used in the fabric over a given period can be given by the consumption in metres of weft/h:

$$P'''_{fL} = 60 \cdot n \cdot i \cdot B_b \cdot \eta_L \qquad [10.11]$$

where B_b = the length of weft threads woven (drawing-in width in reed B_r plus the length of the weft inside the selvedges), m.
For example, for n = 280 rpm; i = 1; B_b = 1.62 m,
we obtain $P'''_{fL} = 60*280*1*1.62*0.88 = 23\,950 \text{ m/h}$

The coefficient of useful time η_L of weaving machine operation is defined by the ratio of actual operational time of the weaving machine for the estimated time in relation to all estimated time. Thus, the η_L depends on weaving machine idle time. Idle time is divided into technical and technological idle time. Technical idle time of a weaving machine occurs because of faults with the weaving machine and the need to carry out repairs. The portion of idle time related to capital repair is not considered when calculating

η_L; these repairs are considered separately in the form of a percentage of planned stoppages. Technological idle time occurs owing to the breakage of warp and weft threads, warp beam installation and its removing, etc.

The greatest influence on idle time is yarn breaks; therefore, to increase the productivity of the weaving machines, it is necessary to improve yarn quality, improve its preparation for weaving, improve the technical performance of weaving machines and set up the optimum drawing-in parameters on the weaving machine. As a result of these measures, the designer-technologist can obtain comprehensive information concerning measures of efficiency in the production of good quality fabric.

10.8 Comparing operating conditions for natural and synthetic fibres

In many cases, woven fabrics formed from natural fibres are intended for use by people and those based on synthetic fibres are intended for technological applications. Weaving machines are designed according to the specific character of the woven fabrics to be produced (Ref. 3). Much of the basic knowledge in the field of weaving was developed through the manufacture of fabrics from natural fibres. Modern synthetic fibres can be much finer than natural fibres. As a consequence, the warping process has to deal with fibres or yarns with novel properties. This introduces certain difficulties into the weaving process. Therefore, to use accumulated knowledge in the weaving of synthetic fibres, we need there to be a suitable *analogy* to the geometrical parameters of natural fibres.

On the basis of this observation, some possible future trends in improving the weaving conditions of flat woven fabric can be given as follows:

- All weaving machine mechanisms need to be controlled by a central control unit (CCU). To achieve this, it is necessary have suitable detectors on all the basic mechanisms.
- The appropriate values of tension and the extension of warp threads and fabric in the ESFF have to be achieved by continual passing of information to the CCU.
- The CCU carries out a comparison of this information with the optimal data and issues appropriate control signals to the relevant mechanism in the weaving machine. The result is the optimal control of the process of woven fabric formation and a stable fabric structure, accompanied by minimal breaking of the warp threads.
- The maximum velocity of the weaving machine drive will be determined by the CCU.

Weaving machine parameters for specific woven fabric structures

- The cloth roller must have a maximum capacity that accords with the capacity of warp beam, so that machine stoppages will be minimized.
- All warp threads in both the top and bottom branch have to be under equal tension during shed formation so as to avoid any problems due to warp stretching.
- For shedding and beating-up mechanisms, it would be advantageous to use tappet devices to maintain the set level of warp tension in front of the fell of woven fabric in the beating-up phase.
- The selvedges of the fabric should have a thickness no greater than the thickness of the fabric.

10.9 Questions for self-assessment

1. What does the concept of 'normalization of the weaving process' mean?
2. What are the causes for the quality of a woven fabric of a given structure to differ between different weaving machines of the same type?
3. What final criteria should be used to estimate the efficiency of normalization of the weaving process?
4. What are the factors that define the efficiency of the fabric formation process?
5. What is the sequence of operations of normalization of the weaving process?
6. What are the parameters of the operation of a weaving machine that are required to be calculated?
7. With what device is it possible to establish the settings of the dropper device and the back-rest on a weaving machine with respect to the neutral line (NL) of the shed?
8. How it is possible to use the method of marks on warp threads to control the efficiency of conditions of fabric formation on a weaving machine?
9. How is it possible to define a method of graphical representation of processing tension on warp threads on a weaving machine?
10. Give examples of extreme and optimum conditions of fabric formation as shown by the profile of an oscillogram of warp thread tension on a weaving machine.
11. How is it possible to make a cyclic diagram of coordination of the operation of the mechanisms of a weaving machine?
12. In what units is it possible to define the productivity of weaving machines?
13. What are the future trends in improving the quality of weaving of flat woven fabrics?

10.10 References

1. Gordeev V.A. and Volkov P.V., 'Weaving', 'Leg. and Pitsh. Prom.', Moscow, 1984 (in Russian).
2. Choogin V.V., Kahramanova L.F. and Nedovisiy M.N., 'Technology of Weaving Manufacture', State Technical University, Kherson, 2008 (in Russian).
3. Choogin V.V. and Chepelyuk E.V., 'The Forecasting of Manufacturability of Woven Fabric Structure', State Technical University, Kherson, 2003 (in Russian).
4. Chepelyuk E.V. and Choogin V.V., 'Friction of Weft on Weaving Machines', Monograph, National Technical University, Kherson, 2008 (in Russian).

11
Control of woven fabric quality: defects and quality assurance of grey fabrics

DOI: 10.1533/9780857097859.171

Abstract: This chapter describes the methods and special equipment used for the evaluation of the quality and the quantity of the woven fabric produced.

Key words: quality of the woven fabric, quantity of the woven fabric.

11.1 Introduction: the quality control of woven fabric

Fabric removed from the weaving machine and awaiting the finishing operations, is referred to as grey (greige) fabric. Efficient control of the production of fabrics requires the availability of adequate information on the sequence of technological processes involved, the operation of the machines involved, and the management of the stock of semi-finished products during all stages of completion and the required warehouse space. The knowledge and the analysis of this information enables the head of the weaving section to estimate and make well-informed decisions on progressing the work of all sections of manufacture (Ref. 1 and Ref. 2).

With the increase of the variety of yarn types and packages, types of machines, articles of fabric, etc., the necessity for quick estimation of material volumes has resulted in the use of computerized automated control systems of weaving manufacture (ACSWM).

To enable the application of ACSWM, each process machine should be provided with appropriate devices for the determination (evaluation) of the following parameters:

- speed of operation;
- extent of short-term idle times due to warp and weft breakages;
- replenishment of yarn packages;
- machine faults;
- machine idle time due to various causes (e.g. cleaning of machines, etc.);
- production volumes achieved.

This production information is automatically recorded by an ACSWM. An operator should enter details of machine idle time due to replenishment of stock, repair, yarn breaks and other such causes into the ACSWM in advance by means of special input panels (e.g. one panel for 16 or more weaving machines). All such information is collected at a rate of some 10 updates per minute, and is registered and displayed on the central control unit by means of a mini-computer (Uster).

The system enables the service engineer, the operator and the section head to evaluate the productivity of each worker, weaving machine or any other equipment over any given period. The head of each technological process can analyze the duration, the variety and the reasons for idle time in sufficient time to be able to take the necessary measures to maintain machine productivity and the quality of the product. The loss in productivity of equipment is often caused by the unavailability of estimates of the correct level of yarn quantity. For example, in the woollen industry this makes it difficult to divide a quantity of a yarn into the warp and the weft for the development of a fabric of a given colour or structure. This may also occur in the development of coloured yarn from cotton.

The most time-consuming technological operation in weaving is the replenishment of warp. Forecasting the exhaustion of warp in a weaving machine involves starting with the installation of the full warp beam on each weaving machine, followed by visual observation over several days. Without ACSWM, the weaving master or other control personnel may find several weaving machines stopping simultaneously, needing the replacement of warp beams (which should be avoided to minimize idle time). Information obtained from ACSWM enables the section head to plan and more accurately coordinate work at all stages of weaving manufacture for the uninterrupted supply of grey materials of each article of fabric. After production on weaving machines, the grey fabric arrives in the quality control department where the quality assessment of the fabric is carried out by manual or mechanical means, and preparation or packing is carried out for transfer of the fabric to the finishing factory or the finished goods warehouse.

11.2 Defects in grey fabric

Various circumstances can cause fabric defects: the unsatisfactory quality of a yarn, or of auxiliary materials; weaving machine faults; lapses of concentration of the weaver or the service engineer etc. Defect detection methods and quality control procedures may vary in different countries. However, the classification of typical faults is basically the same (Ref. 1 and Ref. 2).

In grey fabrics, irrespective of the type of weaving machine on which they are made, the following defects can occur:

- **Missing warp ends**: absence of one or several warp threads at a certain place in the fabric; the reason for this defect being due to warp stop motions with droppers.
- **Warp floats**: defects of interlacing on a small area due to a group of warp threads floating over one or several weft threads (e.g. when the end of a broken warp thread becomes entangled with the adjacent threads); the reason for this fault is also warp stop motions with droppers.
- **Weft stripes of reduced density**: widthwise stripes in a fabric with a lower density of weft (thin places); this fault is caused by a defective warp or cloth regulator.
- **Weft stripes of higher density**: widthwise stripes in a fabric with a higher density of weft (thick places); as with stripes of lower weft density, this defect occurs as a result of faulty warp or cloth regulators.
- **Weft floats**: a group of weft threads not interlacing with the warp over a small area of the fabric and appearing in the form of loops on the face or the back of the fabric; this can be due to: an unclean shed (no clear shed lines), incorrect timing of weft insertion in the shed, sagging of a group of warp threads, a arrangement of temples that is too low or too high.
- **Longitudinal stripes**: warp threads that have not interlaced over some length of the fabric and the formation of floats of different length; the reason for this is the weakened tension of a group of warp threads following breakage removal or wrong drawing-in in the reed.
- **Couples** (reediness): the warp threads are found to be grouped in bunches, leading to visible longitudinal streakiness in the fabric; the reason for this can be the wrong choice of reed, improper drawing-in of threads in the reed dents, or inadequate shed levelling (insufficient imbalance).
- **Drawing-in faults**: broken interlacing along the fabric; this appears as a result of a faulty shedding device, or at start-up of the weaving machine without a broken warp thread having been repaired.
- **Unequal picking**: full width stripes showing a variation in weft density; a principal cause for this is a faulty warp or cloth regulator.
- **Reed marks**: a rarefied (widened) strip along a fabric due to a bent reed wire.
- **Ring temple punctures**: lines formed by punctures along the fabric edge caused by a malfunctioning temple.
- **Weft loops**: Protruding and (more often at selvedges) unstraightened or twisted weft threads; this is caused by excessive twisting, insufficient humidity, or insufficient weft tension.
- **Bad selvedges**: wavy, non-uniform, flabby, or hard, damaged selvedges occur for various reasons: weak or non-uniform tension of the threads at the selvedges; incorrectly chosen interlacing or drawing-in of threads in

heald eyes, reed, or droppers; a faulty selvedge mechanism on shuttleless weaving machines.
- **Start-up marks (set marks)**: stripes extending across the whole width of fabric with reduced or increased weft density may occur in fabrics of low or average density on the start-up of the weaving machine following a stoppage; the reason for this is relaxation in the stretch of warp threads during extended weaving machine idle time. These marks can also be caused by faulty warp beam regulators or cloth regulators.

Apart from the above types of fault, other types of defects may also occur on occasion (e.g. incorrect weft (in shade or count), thick and thin warp threads, damaged threads, pollution (dirt), oil stains.

Each type of weaving machine also has fabric faults peculiar only to that type of machine. For example, in automatic shuttle weaving machines, on the change of a pirn, a drag-in fault can occur due to the previous pick becoming drawn into the shed and being woven into the fabric. The cause of this defect is imbalance in the temple scissors. For fabrics manufactured on air-jet, water-jet, and air-jet rapier-rapier weaving, 'short-weft' can occur. This is an incomplete weft over the width of fabric. Interlacing faults are caused by the end of the weft being dropped in the shed. The end of the weft may also turn back on itself. This defect can be caused by faults in the weft feeder, or excessive or insufficient air pressure.

Centralized control of the weaving room employs the principle of bi-directional communication between the weaving machine and the computer for programming (computer aided design (CAD) and computer aided manufacturing (CAM) systems), and for quality control in real time during production. This type of online control permits immediate intervention by a human operator to remove faults.

11.3 Quality assurance of grey fabric

The quality of grey fabrics is defined according to various state standards and specifications. In many countries, a catalogue of defects published by the Swiss Textile Institute of Zurich, and distributed by the International Textile Service (ITS), of Schlieren, Switzerland is used. For example, in the Ukraine, Russian standards of woven fabric defect are used (Ref. 1 and Ref. 2).

The establishment of the rating of a fabric depends on the fabric group to which it belongs. So, for example, cotton fabrics share four groups:

1. fabrics from combed yarn: plain, sateen, clothing, dress, coloured-yarn, furnishings;
2. grey, gingham, plain, beaver, duffel;

3. fabrics from low grades of cotton for mattresses, pillow cases;
4. fabrics with cut pile.

Sorting of the grey fabric is based on defects of appearance and on indicators of physico-mechanical properties. For example, for grey cotton fabrics in the Ukraine, two grades are used: 1 and 2. Fabric sorting is defined by a score, conditional on the length of a piece by the total number of defects counted on it. The total amount of points for the conditional length (30 m) of a piece is assumed to be: for Grade 1: no more than 10 points; for Grade 2: no more than 30 points. If the total count of points is more than 30, the fabric is not gradable. Defects of appearance are identified by viewing the right (face) side of the fabric on a fabric inspection machine in the weaving mill. The deviations from specifications are decided according to physico-mechanical indicators estimated on the basis of results of laboratory research.

The estimation of the rating of fabrics from other types of fibre involves similar features. So, silk fabrics can be assigned to Grade 1, 2 or 3; linen, to Grade 1 or 2. The total number of points for a rating estimation may similarly vary. For silk fabrics on a piece 80 m in length for Grade 1 (sort), there should be no more than 12 points; for Grade 2, no more than 30 points; for Grade 3, no more than 54 points. In linen manufacture, the rating is defined by the total number of points on an area of fabric 30 m^2.

11.4 Equipment for the control of woven fabric quality

As has been noted above, fabric control is carried out by defect-registration machines (Ref. 1 and Ref. 2). For example, weaving mills widely use type DR-2 (Russian) machines. On these machines (Fig. 11.1(a)) it is possible to change the direction of movement of the fabric. The fabric moves from the cloth roller (1) by means of system of cylinders (2, 3 and 4) on the viewing table (5). The viewing table has a glass base (6) with a backlighting device 7. The angle of slope of the working surface can be changed depending on the relief of the fabric and the height of the operator. Following inspection, the fabric turns around the directing cylinder (8) and remains within the stacker (9) for collection on the table (10). Instead of a spreader, the fabric can be wound on a roll. The speed of movement of the fabric can be set at 16, 30, or 43 m/min. The machine is equipped with a calculator to record the length of all pieces wound. Separate records are made for the output of the weaver for all work, arranged by day. To clean and control the hairiness of fabrics, USB (universal, shearer, bisurface) shearing machines are employed.

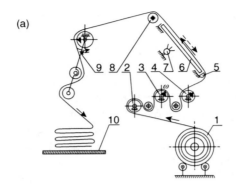

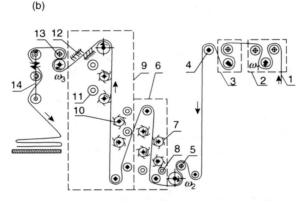

11.1 Scheme of (a) defect-registration machines and (b) scheme of shearer-cleaner machine. Refer to text for detailed explanation of components.

To avoid the occurrence of defects on finishing, and so as to obtain a fabric clear of fluff and shearing, uncut ends of threads, small knots, loops, etc., the USB (Russian) technological system is used, Fig. 11.1(b). The fabric (1) moves to the drawing device of the machine through smoothing out platens (2), tension platens (3) and a fabric-directing platen (4). The fabric then passes through a brake device (5) and arrives in the cleaning chamber (6). Here, the fabric is cleared by means of four brushes (7) on both surfaces. The lint is sucked away by air lines (8). Further, the fabric gets sheared in chamber (9) which has four shearing devices (10). Each device consists of a shearing cylinder with 24 plates and a knife. The shearing chamber is also provided with a suction cleaner (11). The fabric is cleaned by brushes (12) and, by means of pulling platens (13), moves to the stacker (14). The operational speeds of the fabric achieved by a USB machine are 26, 40, and 81 m/min.

Recently, for the purpose of increasing labour productivity and to provide equipment for grading and fabric accounting, control-lines (CL) machines have been adopted. These lines carry out the technological processing of the fabrics and fabric transport functions. There are three kinds of CL machine: continuous action, interrupted (settable) actions and three-branch. Of these, the first are used at factories with a small number of articles of fabric; the second, at factories with a large assortment of fabrics. Lines of the third type are applied to cleaning, grading and quantifying unfinished blended-mixture fabrics.

CONTROL-LINES (CL) (Russian), with continuous action, consists of a store for fabric rolls, unwinding devices, a sewing-machine, two or three quality control-registration tables, interoperational compensators for fabric, measuring-packing machines, or winding devices. For the quality control of cotton, linen and silk fabrics, 'Stema' product lines are used.

Apart from production control, and the accounting of the rating and quantity of a fabric, modern machines also clear the loose ends of threads, and clear oil stains by means of a detergent liquid and aspirator pistol-spray is also carried out. Continuous counting of all defects according to their type and disposition, and the width of fabric completely characterize the quality of a fabric.

For the quality control of expensive fabrics (e.g. woollen types, fabrics with strongly pronounced patterns of thread interlacing, fabric with jacquard drawing), manual restoration of interlacing of the fabric is carried out. Good quality assurance and production accounting of fabric demands an increased number of workers. For the automatic grading of a fabric at increased speeds (up to 800 m/min), laser-based systems such as the Sick-Scan-System Ko-Re-Tra (Germany) can be used.

11.5 Questions for self-assessment

1. What are the parameters required to be known to control machines related to weaving production?
2. How is it possible to automate a system of control of parameters for the operation of weaving machines and the prevention of defects of fabric?
3. What are the causes of formation of the different types of defects in a woven fabric?
4. Name the types of woven fabric defects which have the highest possibility of occurring?
5. What are the specific defects of fabric which are likely to occur on different kinds of weaving machines?
6. How is the identifying and counting of fabric defects carried out?
7. What is the nature of equipment necessary for identifying and counting defects in a fabric?

8. What are the basic devices required on a defect-registration machine for visual quality assurance of a fabric?
9. What are the operations carried out on a shearer-cleaner machine?

11.6 References

1. Gordeev V.A. and Volkov P.V., 'Weaving', 'Leg. and Pitsh. Prom.', Moscow, 1984 (in Russian).
2. Choogin V.V., Kahramanova L.F. and Nedovisiy M.N., 'Technology of Weaving Manufacture', State Technical University, Kherson, 2008 (in Russian).

12
Movement of raw materials and finished fabrics in weaving manufacture

DOI: 10.1533/9780857097859.179

Abstract: This chapter describes the use of electrically driven vehicles for the transportation of raw materials and outputs between different units of production at the weaving factory.

Key words: electrically driven vehicles, chain floor, floor magnetic conveyance, suspended conveyor.

12.1 Introduction: product transportation in weaving manufacture

In weaving manufacture, yarn supplies arrive in a variety of package types depending on their raw structure and configuration: spinning pirns, bobbins, skeins, cones, etc. If the weaving department is in a separate building at a considerable distance from the spinning mill or synthetic fibre factory, yarn packages can be moved by road or rail (Ref. 1 and Ref. 2). Almost invariably, such commodity yarn is packed into bags or pallets of considerable weight and volume. Such packages are unloaded and transferred to the warehouse on racks, together with relevant information on the type and quantity of the yarn involved, by means of an electroloader. If weaving manufacture is a component of a textile industrial complex, yarn packages (on pirns, bobbins, etc.) can be transferred to the central warehouse on special carts with spikes ('fir-trees'), or in bulk in carts or in boxes, each holding between 40 and 100 bobbins. The carts can be moved by electrotractors, by chain floor or suspended conveyors.

12.2 Transportation of raw materials and outputs

From the central warehouse, the yarn is transported either into the winding section, or directly into the warping department, by means of an electrotractor, a chain floor or suspended conveyor, floor magnetic conveyance, etc. Over short distances, transportation of the yarn by warehouse operatives is

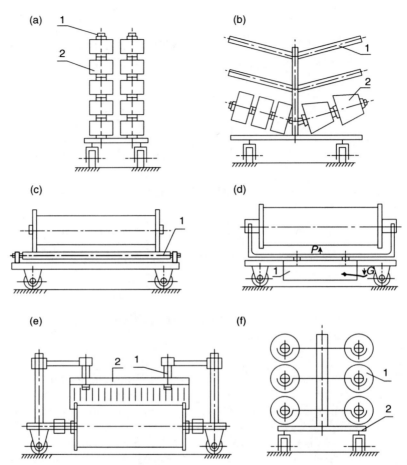

12.1 Means of transportation of the yarn and the fabric. (a), (b) Spike carts for bobbins; (c), (d) carts for warp beams; (e) cart for the drawing-in of warp; (f) cart for woven fabric rolls. Refer to text for detailed explanation of components.

often more economical. Following rewinding, if this has been necessary, the yarn can be returned for storage in the central warehouse, or moved into an intermediate warehouse before warping. Moving the packages (weft pirns, bobbins) can be carried out by belt conveyors that automatically weigh and record the identity of the car and its operative. The bobbins are frequently placed directly at the creel of the warping machine (Fig. 12.1(a) and (b)).

After the warper's beams have been prepared, they are usually moved to the sizing department on a pendant monorail electric drive, where they are placed near the sizing machines. On section warping machines, the warp beams can be also be moved into the drawing department by a pendant monorail electric drive, or on floor carts by an electrotractor. The same type of transport is used for moving warp beams following sizing.

In the weaving department, warp beams are transported on floor carts (Fig. 12.1(c), (d) and (e)) by means of an electrotractor. These carts can be provided with basic rollers (1) (Fig. 12.1(c)) or with a hydraulic lift (1) (Fig. 12.1(d)), operated by pedal G. For transportation of warp beams, together with thread drawing-in devices (2) (harness, reed, droppers) Fig. 12.1(e), carts with arms (1) are used. The weft yarn from a warehouse is transported to the weaving department on carts (Fig. 12.1(a)) by an electrotractor. The cart has only four spikes of moderate dimensions, and can be accommodated without causing hindrance in the weaver's shop. In the woven fabric quality control department, the fabric rolls are moved by electrotractors on carts provided with horizontal hooks (Fig. 12.1(f)). The width dimensions of electrotractors should not exceed 550 mm, and the speed of their movement in processing shops should not exceed pedestrian speed.

12.3 Questions for self-assessment

1. What are the different types of yarn packages that arrive at a weaving mill? How are they packed during transport?
2. What are the different types of moving machinery that can be used to transport the yarn to the central warehouse of a weaving mill?
3. What type of transport is used to move yarn to the weaver's shop?
4. What type of transport is used to take grey fabric from the weaver's shop to the quality control department?

12.4 References

1. Gordeev V.A. and Volkov P.V., 'Weaving', 'Leg. and Pitsh. Prom.', Moscow, 1984 (in Russian).
2. Choogin V.V., Kahramanova L.F. and Nedovisiy M.N., 'Technology of Weaving Manufacture', State Technical University, Kherson, 2008 (in Russian).

Appendix 1: Further reading on weaving technology

This bibliography covers various theoretical and practical aspects of weaving technology relevant for students and teachers at textile teaching establishments. Since training is usually of limited duration, only a select number of published works in the field of weaving is included. It is necessary to develop an adequate level of knowledge to start working with confidence after the completion of training at an institute. In the course of their training, we recommend that students follow not only the specified textbooks, but also those texts describing the principles and mechanisms of weaving machines and their practical use. There is one fact which it is very important to remember. Many variants of each mechanism have been developed over time. We have included works which describe the basic mechanisms of weaving machines which have remained unchanged. This fundamental information remains valuable in understanding the nature of different mechanisms, to compare different devices and to use them effectively in practice.

Bibliography

Adanur S., *Handbook of Weaving*, CRC Press, 2001 (in English).
Choogin V.V., Kahramanova, L.F. and Nedovisiy M.N., 'Technology of Weaving Manufacture','State Technical University', Kherson, 2008 (in Russian).
Fox T.W., *The Mechanism of Weaving*, Macmillan & Co. London and New York, 1894.
Ghandi K. (ed.), *Woven Textiles: Principles, Technologies and Applications*, Woodhead Publishing Limited, 2012 (in English).
Gordeev V.A. and Volkov P.V., 'Weaving', 'Leg. and Pitsh. Prom.', Moscow, 1984 (in Russian).
Greenwood K., 'The Control of Fabric Structure', Merrow Publishing, 1975 (in English).
Kuligin A.V., 'Automatic Weaving Machine AT-100–5M', 'Iv. Knig. Publ.', Ivanovo, 1959 (in Russian).
Kusovkin K.C., Danilov V.V., Kurochkin V.N., *et al.*, 'Experience of work by weaving machine STB (Sulzer)', 'Leg. Ind.', Moscow, 1968 (in Russian).

Lapisov A.G., 'Directions on the analysis of woven fabrics and elaboration of the weaving plan', 'Tip.Orgelbranda', Varshava, 1903 (in Russian).

Lord P.R. and Mohamed M.H., 'Weaving, Conversion of Yarn to Fabric', Woodhead Publishing, 1999 (in English).

Malishev A.P. and Vorobiev P.A., 'Mechanics and Calculations of Weaving Machine Construction', 'Mashgis', Moscow, 1960 (in Russian).

Marks R. and Robinson A.T.C., 'Principles of Weaving', Textile Institute, 1976 (in English).

Mitaev G.F. and Panov V.A., 'Mechanisms and Set-up of Pneumatic Weaving Machines', 'Legprombitisdat', Moscow, 1987 (in Russian).

Ormerod A., 'Modern Preparation and Weaving Machinery', Butterworths, 1983 (in English).

Ormerod A. and Sondhelm W., *Weaving: Technology and Operations*, Textile Institute, 1995 (in English).

Rosanov F.M., Kutepov O.C., Gupikova D.M. and Molchanov S.V., 'Structure and Practice of Woven Fabric', 'Legprom', Moscow, 1953 (in Russian).

Schwartz. P., Rhodes T. and Mohamed M.H., 'Fabric Forming Systems', Noyes Publications, 1982 (in English).

Stepanov G.B., 'Weaving Machines ATPR: Structure and Parameter Calculations', 'Leg. and Pish. Prom.', Moscow, 1983 (in Russian).

Talavasek O. and Svaty V., 'Shuttleless Weaving Machines', 'Legprombitisdat', Moscow, 1985 (in Russian).

Vasilchenko V.N., 'Beating-up of Weft Thread', 'Legprombitisdat', Moscow, 1993 (in Russian).

Vasilchenko V.N., 'Investigation of Beating-up Process of the Weft Thread', 'Legpromisdat', Moscow, 1959 (in Russian).

Appendix 2: Glossary of terms applied to weaving machines and weaving technology

Chapter 1

Back-rest – The cylindrical bar or roller over which the warp threads are passed upwards from the warp beam and then horizontally on to the weaving machine; the back-rest also forms part of the warp let-off mechanism so as to maintain warp tension at a constant mean level during weaving.

Beating-up of the weft – The action of the reed which consolidates the newly-inserted weft thread into the cloth-fell.

Breast beam – The rigid cylindrical bar or tube which sets the length of fabric produced in the fabric forming zone and which changes the direction of the fabric from a horizontal to a vertical plane on the weaving machine.

Cloth roller – The cylindrical beam or tube onto which the taken-up fabric is wound.

Dropper – The steel plate-gauge for detecting the integrity of an individual warp thread on the weaving machine (i.e. warp break detection).

Dropper device – Arrangement which operates with the droppers to stop the weaving machine in the event of warp thread breakage.

Elastic system of fabric formation (ESFF) – The sections of the warp threads from the warp beam to the fell of the fabric, and the fabric from the fell to the emery roller, that are subject to elastic deformation and fluctuation of the tension of the warp threads and the fabric.

Emery roller – The rigid cylindrical beam or tube with a rough surface (to afford frictional contact to prevent slippage) for the withdrawal of fabric from the fabric-forming zone prior to its being wound onto the cloth roller.

Fell of the woven fabric – Edge of woven fabric on the weaving machine as defined by the last pick (weft) beaten-up into the fabric.

Fabric formation zone – On the weaving machine, the area of the fabric between the newly inserted pick and the last pick in the cloth-fell which has been consolidated into the fabric.

Harness – The assembly of frames which carry the heald wires to form the shed from groups of warp threads on a weaving machine.

Machine-yarn-fabric path (MYFP) – The path of warp threads on a weaving machine from the warp beam to the fell of fabric, and the path of the woven fabric from the fell to the cloth roller.

Picking of the weft – The insertion of the weft yarn in the shed by the weft insertion device.

Preparatory section of the weaving mill – The section of the mill where warp and weft threads are treated (e.g. cleared, sized, etc.) and wound into the appropriate packages ready for use on weaving machines.

Reed – A frame with steel wires or plates that evenly distributes of warp threads across the width of the fabric, and which also beats up the weft. It may also form part of the movement of the picking device in the shed.

Reed blades – Reed elements in the form of profiled plates.

Reed split (dent) – Space between two adjacent reed blades for warp thread setting (for the uniform spacing of warp threads).

Shed – Space formed between the top and bottom layers of warp threads when they are opened for the picking of the weft.

Take-up roller – The cylindrical bar used to maintain a firm grip on the cloth as it is produced and withdraw it at a constant rate following each beating-up.

Temples – Devices to maintain the fabric at the cloth-fell at the same width as that of the warp in the reed.

Thread (Yarn) – A thread consists of twisted strands of short or continuous fibres.

Warp beam – The rigid tube or beam with rigid flanges on which a parallel group of threads (600–3000) can be wound for supplying as warp on a weaving machine.

Warp threads – The longitudinal system of threads on a weaving machine or in a woven fabric.

Weaving – Process of forming a fabric using a system of longitudinal (warp) threads by the successive addition of a system of cross (weft) threads, so as to provide a distinct fabric structure.

Weaving machine – Machine, typically power driven, used to produce woven fabric.

Weft threads – System of threads running across a fabric.

Yarn – Threads made by the suitable twisting of natural fibres, or continuous stapled filaments (15–50 mm in length) of synthetic fibres.

Chapter 2

Delivering (or feeding) of warp threads – method of releasing warp into the weaving machine by the regulated unwinding of the warp beam, rather than the unwinding action being initiated by the *tension* of the warp yarns in the shed.

Differential lever – A lever on the friction warp brake for the automatic regulation of brake band tension during the gradual reduction of the radius of the warp beam.

Manufacturability – This is the optimal degree to which threads are subject to deformation during fabric formation on a weaving machine.

Primary setting of warp threads – This is the optimum *minimal* level of the primary setting of the warp tension during fabric formation in the weaving cycle.

Release of warp threads – The length of warp released from the warp beam under the required *tension* of the warp threads. The length warp released will be slightly more than the length of the fabric cell because warp threads are crimped in the fabric forming zone.

Stabilization of the mode of release and the tensioning of warp threads – The warp beam should unwind so as to maintain warp tension in the weaving machine at a constant average level. Typically, two conditions should be satisfied. The length of warp released should ensure that the released warp is under a constant tension and the back-rest fluctuates cyclically at a set level (range).

Warp beam brake – This enables the intermittent *controlled release* of warp threads by means of *passive (frictional) resistance* to the turning moment relative to the axis of the warp beam and functions by virtue of warp tension.

Warp regulator – This provides for the *controlled release* of a length of warp threads so as to manage the warp tension in the ESFF of the weaving machine; it serves to control the magnitude of warp release, and to provide *active resistance* by means of a mechanism (a worm gear, planetary gearing, a belt variator, hydraulic system, differential frictional system, electromagnetic system, etc.); it is driven independent of the main shaft of the weaving machine.

Chapter 3

Bottom half-shed – The shed geometry obtained by lowering some of the warp threads and leaving the rest remaining in the NL (neutral line) position.

Comber board (Jacquard machine) – A board with guide eyes which determines the array of harness cords.

Cyclogram of shedding – succession of shedding movements undergone by each warp thread in one weave repeat. For example, when the plain weave (shedding cycle 1/1) is produced, one-half of the warp threads moves up to the top shed and the other half moves down to the bottom shed; this allows one weft thread to be inserted during one rotation of the main shaft (0°–360°). During the second rotation of the main shaft (360°–720°), these two groups of threads exchange position and the next weft is inserted. This cycle repeats as weaving continues.

Dobby – The mechanism used to control the shedding process on a weaving machine for repeat interlacings involving up to 24 weft threads.

Dwelling phase – The range (between 90° and 120°) between which warp threads are in a fully open shed.

Even shed – Warp threads in the front part of each shed line ensure that the shed line will be in the same plane; this improves the picking of the weft thread in the shed.

Filling design pattern (weaving plan) – The weaving plan consists of four elements: repeats of the interlacing of warp and weft threads (fabric ground and selvedges), the draught plan in the reed, the draught plan of the heald, and the system by which the threads are drawn into the droppers.

Griffe block (Jacquard machine) – A lifting bar with plates (knives) to lift the hooks on shedding.

Harness cord (Jacquard machine) – A cord that joins a heddle to a neck cord.

Heddles (Jacquard machine) – Typically made of cord or wire, each heddle has an eye through which an individual warp thread is passed. Heddles perform the same function as the harnesses on tappet and dobby mechanisms.

Heald hook (Jacquard machine) – Hook, which is attached to neck cord and is controlled by needle knees.

Jacquard machine – A device that forms a shed from small groups of, or even individual, warp threads (maximum: 2600 small groups or separate threads). The Jacquard machine only lifts warp threads; this is achieved by means of a special prism and punched cards. The warp threads are lowered under gravity by lingos or rubber tension bars.

Knife of dobby mechanism – A bar with a special profile for moving the lifting-hooks on a dobby.

Leasing device – This consists of two dividing rods rocked by a three-armed (or two-armed) levered cam. The rods divide the warp threads into two layers, and can vary the stretch the top branch of the shed more from that of the bottom branch (so as to give an 'unbalanced' shed).

Level phase of the warp threads – This is the degree (20°, ... , 25°) of rotation of the main shaft when the warp *threads are briefly motionless* (*since the heddle eyes are longer than the diameter of warp threads*) at the mid-point of the shed while *heddle movement* is in progress.

Lingos (Jacquard machine) – Lingos are weights used to tension the harness cords of a Jacquard machine.

Mixed shed – A mixed shed is obtained when one of the shed lines (most frequently, the bottom shed line) is set up as an even shed and the other is set up as an uneven shed.

Neck cord (Jacquard machine) – The cord for attaching a harness cord and a heel hook.

Needle boards (Jacquard machine) – Two perforated boards for the array of horizontal needles with knees (to deflect the hooks).

Needle device (Verdol Jacquard machine) – This device is located between the prism carrying the continuous card chain and the main needles of the Jacquard machine. It reduces the force exerted by the needles on the cards and, hence, extends their life.

Needle knees (Jacquard machine) – Needle knees are bends formed on horizontal needles which move the heald hooks under the control of punched cards.

Neutral line of the shed (NL) – This connects the fell of the woven fabric and the centre of the eye of the heald shaft at the mid-position of the harness.

Phase of shedding – This is either when the shed remains in a fixed position, or the movement of the warp threads over a specified section of the shedding curve, usually expressed by the angles of rotation of the main shaft.

Phases of shedding – These are the angles of rotation of the main shaft when the warp threads are in the level phase or are being positioned to form the shed.

Prism of dobby mechanism – A roller, with grooves for shedding cards.

Punched cards (Jacquard machine) – The cards control the sequence of the lifting of the warp thread in the shedding process.

Rotation of tappet of shedding device – The number of rotations of the main shaft for *one* rotation of the tappet.

Shed closing phase – The degree through which warp threads are moved (90°, ... , 130°) from the dwelling position to the level position.

Shed opening phase – The degree through which warp threads are moved (90°, ... , 130°) from the mid-point to the fully open shed.

Tappet shedding device – A mechanism for shed formation controlled by tappets, driving between 2 and 14 harnesses. The tappets are closed either by the activation of a spring on the treadle levers, or by grooved tappets or double tappets.

Top half-shed – This is obtained by raising some warp threads to their top position and leaving the rest in the NL position.

Trap (or collar) board (Jacquard machine) – A board with openings for the array of neck cords.

Uneven shed – This takes place when the harnesses move a similar distance, but the threads in the front part of the shed are located in a different plane.

Variable tension (or unbalanced) shed – There are various ways to achieve this. When the dropper bar is raised higher than the NL, the bottom (lower) branch of the shed is stretched more, creating a *positive variable tension* of the shed branches. When the dropper bar is lowered, the top (upper) branch is stretched more than the bottom (lower) one, creating a *negative variable tension* of the shed branches. It is possible to have an *equally tensioned* (or balanced) shed, by adjusting the height of the dropper bars.

Chapter 4

Automatic change of weft packages – This operation can be carried out in both shuttle and shuttleless machines. Four devices are required to achieve the automatic change of weft packages: a gauge detecting the presence/depletion of weft, the change mechanism, a supply of weft, and a safety device.

Automatic mechanism for shuttle change – This provides for the automatic change of shuttles with pirns (for yarn of a different colour, or different type of fibre etc.) without the need to stop the weaving machine. This mechanism ejects the spent shuttle by means of a lever.

Battery type weft supply – On shuttle machines, a range of pirns of six kinds is used: round (a capacity of 28 pirns), tape (12 pirns), feeding hopper (40 pirns), vertical magazine (85 pirns), box or tape (160 pirns), in combination with a moving head, Unifil (5 pirns).

Compensators of weft threads on shuttleless weaving machine – These are used to ease unwinding of the thread from a package (bobbin) in shuttleless weaving machines before the insertion of the weft into the shed. It is necessary to form a loop to maintain a constancy of the tension in the phase of picking of the weft. This process is made by a special lever set in motion by a cam (or by an air draft). The law of movement of the compensator, defined by the profile of a cam, corresponds to the manner of inserting weft on the specific type of weaving machine.

Devices for measuring weft on shuttleless weaving machines – These devices combine the processes of the measurement and accumulation of weft during its continuous take up from the bobbin. There is a more rational variant that maintains constant weft tension in the picking phase by the continuous winding off of the thread from the bobbin and an intermediate measuring drum with a smooth surface from which weft can be drawn off without any harmful resistance caused by the pulling effort of a weft picking device.

Gauges for detecting the presence of the weft on weaving machines – These are activated in two situations: in the event of a weft break, or on the depletion of weft in the shuttle or bobbin in a shuttleless machine.

Hollow (coreless) pirn – A tubular hollow pirn, used in the weaving of rough, thick linen and jute yarns; thread take-up is from the inside.

Manual change of weft packages – This operation is carried out by weavers, usually without stopping the weaving machine.

Multishuttle mechanisms – These are necessary for the development of fabrics with weft of various colours, different fibrous structure or non-uniform thickness (natural silk, yarn from a carded spinning method, etc.). A multishuttle mechanism consists of one or two magazines with shuttle boxes for several shuttles located on one or both sides of the sley beam; it moves relative to the sley beam in a forwards or rotary (the 'revolving' type) movement of boxes operated by a mechanism controlled by means of a punched card 'programme'.

Safety devices of automatic weft supply (pirn change) – In mechanisms that change pirns, there are two kinds of safety device: devices to prevent insertion of the wrong pirn into the shuttle, and scissors (weft cutters) to remove the weft ends that occur during a pirn change in order to avoid flaws in the fabric.

Supply of weft on the weaving machine – On shuttle machines, the supplying of weft relates to both the manual and automatic change of weft packages; weft supply on shuttleless machines is always automatic.

Transfer hammer – This is a basic component of the mechanism for changing pirns: in dependent movement, the transfer hammer is operated from the sley; in independent movement, the transfer hammer is operated from the main shaft.

Weft feeler – The gauge which detects the presence of weft on the pirn in a shuttle signals depletion by various means: mechanical, electric, electroinductive, photoelectric, and also by constant or periodic actions. The mechanical feelers monitor yarn on the pirn by various methods: sliding along a contact surface, penetration of a layer of yarn, the moving apart of weft coils, measuring the diameter of weft winding.

Weft fork – The weft fork detects the presence of weft in the shed following the picking action. In shuttle weaving, mechanical weft forks are are positioned either at the edge of the shed, or inside the shed. *Lateral* weft forks (placed on one side of the warp) detect the presence of weft in the shed following insertion of the shuttle into the shed *but only after two picking operations*. This can lead to the weaving defect called 'shed with missing weft'. *Central* weft forks detect *every* weft inserted into the shed.

Weft thread compensators on shuttleless weaving machine – These are used to ease the unwinding of the thread from a package (bobbin) in shuttleless weaving machines before the insertion of the weft into the shed. It is necessary to form a loop to maintain constant tension during weft picking. This process is achieved with a special lever set in motion by a cam (or by air pressure). The design of a cam varies according to the manner in which weft is inserted on a particular weaving machine.

Chapter 5

Continuous method of inserting weft threads by multiple rapiers – This is provided by moving rapiers set in the lateral grooves of shedding and beating disks. Beating disks press every weft in the fell of the fabric, which is taken up by a pair of drawing rollers and delivered to the take-up roller.

Continuous weft insertion by shuttle – This method is used on circular weaving machines. The movement of shuttles in the shed is carried out in several ways: mechanically (by frictional contact of a shuttle and rubber rollers), by means of an electromagnetic field (where there is no mechanical contact with the shuttle), or by an electric motor in the shuttle. Continuous insertion of weft in the shed is accompanied by continuous processes of shedding and insertion of weft into the fell of the woven fabric.

Continuous weft insertion in the shed – This process of picking the weft on both shuttle and shuttleless weaving machines, which takes place over the full rotation of the main shaft ($0°–360°$).

Continuous woven fabric formation – In this method, the shuttle works in a continuous circular movement on a closed trajectory at a constant speed; this method requires a 'point' beating-up system.

Crank type picking devices – Instead of cams or springs, a system of levers driven by a crank on the main shaft is used.

Force of picking – This is understood as being the amount of kinetic energy given to the picking device (by shuttle or by inertial projectile) in the course of acceleration.

Hydraulic method of weft insertion into the shed – This method is identical to basic system of pneumatic weft insertion. The water stream from a nozzle begins to disperse at a distance of 15 cm, and becomes a mist at 50 cm. The wet fabric is dried in the weaving machine either by a squeezing shaft or by partial vacuum through a slit in the hollow breast beam.

Inertial method of weft insertion – The thread is delivered by means of two contra-rotating cylinders. Through the action of inertia, depending on its rigidity, a weft thread of flax, jute, etc. can travel a certain distance through an open shed.

Inertially small picking device (the projectile) – The ISPD is a device for the inertial picking of weft without a moving store of weft and without a constant mechanical linkage to a drive. As the picking device, spring-based (either by turning a torsion bar by 27°, … , 32°, or with the use of preloaded spiral spring), pneumatic, or electromagnetic methods can be used.

Intermittent weft thread insertion by rapier – This is achieved by picking a thread either in the form of a loop or by its tip. There are various rapier systems: two rapiers enter the shed from opposite ends and the weft is transferred from one to the other when they meet., or a rapier travels the full width of the warp array, or picking in the shed takes place in one or both directions. Rapiers can be rigid or flexible. Rigid rapiers can be in the form of tubes, hollow cores, of constant or variable (telescopic) length. Flexible rapiers can be in the form of punched tapes, reeled up out of a shed on disks or segments, or unilateral or bilateral rapier mechanisms depending on the arrangement of the shed on the weaving machine.

Microshuttle – This is a device for inserting weft threads and, by virtue of its size, has a limited capacity of one pick: it is used in multiphase flat weaving machines. Insertion is microshuttles in the shed is carried out by profiled levers, or press rollers, or an electromagnetic running field along the cloth-fell.

Periodic weft insertion in the shed – This is the picking of the weft on the shuttle or shuttleless weaving machines that occupies only a part of one rotation of the main shaft (90°, … , 220°).

Picking devices of shuttle weaving machines – Movement of the shuttle is provided by a variety of means: a picking stick, tappet, spring, crank, pneumatic or other mechanisms. There can be consecutive and optional picking.

Plate-limiters on an air-jet in shed – This is a series of profiled plates with an internal conical surface that form a channel through which the air from the pressurized air unit mounted on the sley bar flows, and increases the length across which weft thread in a shed can be picked.

Pneumatic (air-jet) weft insertion in shed – This takes place using a two-part nozzle. The weft is delivered in the main nozzle by means of compressed air, following which the compressed air is diffused into the outer nozzle, comes into the shed, grasps the end of the thread and picks it into the channel, as dictated by profiled plates or a profiled reed.

Pneumatic picking device – The shuttle in a pneumatic picking device can be moved in a variety of ways: by a picker, by picking stick, or by a piston operated by compressed air supplied to the cylinder through a distributor.

Pneumatic-rapier weft insertion device – There is no contact between the rapiers in the shed when weft is transferred, and there is no dispersion of the head of air pressure as the flow of air in a rapier tube allows the level of air consumption to be decreased. The major part of the work of transferring weft is executed by the left-hand rapier by means of suction; the right-hand rapier provides a stream of pressurized air. The passage of the rapiers is controlled by a planetary mechanism with external gearing.

Pneumatic-rapier weft insertion – This is the system of weft insertion using two movable hollow 'half' tubes by use of air suction. The tubes conserve air flow.

Positive movement of the shuttle in the shed – Moving the shuttle in ribbon weaving machines by means of toothed gearing (instead of the traditional free flight).

Rapier – This is a device for inserting weft, for either intermittent or continuous methods of weft insertion, without using a moving weft package or pirn; it has constant kinematic communication with its drive.

Relay nozzles on the pneumatic weaving machine – These are used to increase the length of picking weft in a shed and are positioned on the sley at 500 mm intervals.

Shuttle – Device for inserting weft threads in the shed; it carries within its body a stock of yarn for a considerable number of picks. The traditional shuttle makes back and forth inertial movement through the warp shed by means of the energy imparted to it by a 'picking' mechanism.

Shuttles for automatic weaving machines – These have a spring clip type pirn holder and are open at the top and bottom of its body for the automatic insertion of the pirn and its ejection when empty.

Shuttles for mechanical weaving machines – These normally have a wooden body which houses the pirn, and a ceramic eye to allow drawing the weft out, and provided with hardened tips at each end of its body for smooth passage through the warp shed.

Spring picking device – This is a device by which the effort of a stretched spring is applied to the picking stick. The speed of the initial stretching of the spring and the speed of the weaving machine is irrelevant, since it is necessary to provide only the appropriate amount of stretch before picking takes place.

Weft insertion by fluid stream – This is a system where a flow of particles of gas, liquid or solid substances passes the weft thread through the warp shed.

Weave shedding – A weave shed is formed by the consistent movement of the warp threads up and down by means of a harness.

Weft insertion by an electromagnetic field – The picking device in this system travels by means of a running magnetic field created on a linear asynchronous motor on the sley bar.

Chapter 6

Autoregulation of the ESFF – In the event that tension increases in the ESFF, there is a corresponding reduction in the width of the weft strip (cloth-fell length). As soon as the width of the weft strip decreases, there will be also a decrease in deformation of the stretching of warp threads in the beating-up phase, which will lead to a reduction in the level of warp tension in the ESFF. Thus, *without* intervention of the operator of the weaving machine, there will be a gradual process of self-alignment (autoregulation) of tension of the ESFF.

Beating-up of weft into the cloth-fell – This is the method by which a woven fabric is formed and is achieved by the rapid application of force by the reed to the weft to move it into the cloth-fell.

Cell pattern repeats – This is the combination of one warp thread with a particular number of weft threads as determined by the yarn interlacing required by a particular fabric structure.

Condensing of weft in the cloth-fell – This operation can be used in certain fabric-forming operations. It involves moving the newly-inserted weft into the cloth-fell by means of the crossing action of the warp threads, whose tension causes the weft to move into the cloth-fell. (A fabric formed in this manner cannot be woven as closely as a fabric created using the beating-up process.)

Criterion of the optimality of adjusting a weaving machine – This is given by the length of cloth-fell of the given structure of a woven fabric.

Embedding of weft into cloth-fell – This is the continual operation of forming a fabric by pressing the blades of a reed (or laying needles, sliding disks, etc.) on the weft until the point at which that weft is in place in the cloth-fell, and the subsequent crossing of warp threads in the course of formation of the next shed.

Extent of the zone of fabric formation – This is characterized by the quantity of displaced weft yarns (which undergo partial slippage after each beat-up operation (2, ... , 10 weft threads). The actual number of weft yarns which undergo this slippage depends on the nature of the fibre involved, the structures of threads and the woven structure being produced. In this situation, when the next pick is beaten up, the innermost weft which slipped back becomes fully beaten-up and settles down in the cloth-fell.

Fabric cell – This area encompasses the length of warp threads between two successive crossing points enclosing one weft thread.

Fabric-forming components – Various items fall into this category: a beating-up comb or reed, sliding comb, brushes, plates, fabric-forming disks, a crown and a concentrator, depending on the weaving machine concerned.

Fabric formation zone (ZF) – This expression identifies the strip of fabric at the cloth-fell, where weft yarns are moved forwards along warp threads with which they are interlaced during the beating-up operation of the reed.

Formation of the fabric with levelling – The level phase must be completed before the beating-up phase can take place, when the warp threads assume an 'average' height position at beating-up ('early shedding'). In these circumstances, is possible to form a fabric with a higher density of weft.

Formation of the fabric without levelling – This is the beating-up of the weft in an open shed or when the warp threads are in the level phase with a low degree of resistance from warp threads ('late shedding').

'Full-width' beating-up – This is used to join the weft to the fell simultaneously across the full width of the woven fabric.

Heavy beating-up – On a weaving machine with a special reed motion, the reed moves up after the insertion of three or four wefts into the fabric (with the shed changing after each such weft insertion) and forms a loop or pile (looped) fabric (e.g. in towel weaving).

Light beating-up – In this process, the reed does not move up with weft to the cloth-fell; the fabric take-up is greater than the distance the weft is displaced by the reed. In this case, no cloth-fell is formed.

Method of inclusion of the weft thread in an open shed – This method forms a fabric with reduced resistance in the warp threads and helps to prevent weft thread withdrawal (slippage) from the cloth-fell. This is achieved by the prolonged contact of the reed with the weft before it is secured in the cloth-fell by crossing the warp threads of the next shed.

Optimum angle of beating-up – The angle between the plane of the reed and the fabric at which there is no fell sliding along the height of reed teeth. The optimum value of the angle of beating-up at which the force of beating up is directed along a bisector, enclosed between the fabric and continuation of the plane of the lower tensioned branch.

'Point beat-up method' – This is used for the progressive integration of the weft thread into the cloth-fell and is achieved by pressing a short length of weft into the fell using a local fabric-forming component (plates, disks, etc.). The weft is progressively beaten-up along its length until the full length of the weft has been beaten up.

Race board – This serves as one of the guides (bottom) for shuttle movement in the shed.

Ring temples – This special auxiliary device uses an arrangement of rings with needles mounted on an axis to counteract fabric contraction due to crimp formation in the weft at beat-up.

Size of levelling (also called 'shed timing') – This is the distance between the reed and cloth-fell during the 'moment of shed levelling' (of the point at which the formation of a new shed begins). The degree of levelling is established within 20–100 mm during the rotation of the main shaft of the weaving machine by $10°, \ldots, 60°$.

Sley mechanism – This consists of a drive (crank or tappet) and a beam, representing the base or carrier for fastening the fabric-forming component (the reed), which may also guide the weft insertion device.

Volume filling of yarn by yarn (H_V) – This is a complex indicator of the structure of a woven fabric, and depends on the complexity of the process of formation of an element of the fabric. It gauges the thickness and nature of the fibre of the threads, the density of the warp and the weft in the fabric, the shrinkage of threads and the thickness of the fabric.

Weft strip (WS) (cloth-fell length) – The distance of displacement of the cloth-fell by the reed during its movement to its extreme forward position (front dead centre).

Work of stretch deformation of warp threads – A sophisticated indicator of the complexity of the technological process of weaving and the complexity of damaging influences on a warp thread.

Chapter 7

Take-up mechanism – This is used to create the required density of weft in the woven fabric, to take-up the fabric from the fabric formation zone and to wind it onto the cloth roller. Cloth beam regulators may provide continuous or periodic take-up of the fabric.

Take-up of the fabric – Woven fabric is withdrawn from the fabric formation zone on the weaving machine by the take-up motion (cloth regulator). This is achieved with a positive regulator (which is independent of fabric tension, or a negative regulator (which responds to fabric tension), or with a programmed mechanism.

Uniform beating-up of weft threads in the fell of a woven fabric – Under this arrangement, fabric removal from the fabric formation zone is carried out at a rate which depends on the thickness of the weft.

Uniform distribution of weft threads in woven fabric – It is possible to achieve a uniform distribution of weft if the diameter of the weft thread is constant. If the diameter of the weft thread is not constant, spacing between weft yarns will be uneven.

Chapter 8

Collapsible reed – This signals the weaving machine to stop in the event of shuttle problems in the shed.

Controller of weft picking device – This detects of the arrival of a weft projectile in the receiving box In the event that the projectile stops in the shed, a feeler operated by a lever and a spring drops and gives a signal to stop the weaving machine.

Dropperless (contactless) warp stop motion – This system uses a carrier along the two edges of the warp threads (before and after the heald device). One beam checks the integrity of the forward part of the shed between the reed and harness; the other beam checks the integrity of the back part of the shed. In the event of warp breakage, the broken thread sags and crosses the beam during the detection phase, at which point the weaving machine is signalled to stop.

Electric warp stop motion – This is applied on shuttleless weaving machines. In the event of warp thread breakage, the relevant dropper falls and closes an electrical circuit by bridging the external and internal electrodes, both of which are electrically isolated by an insulator. The weaving machine stops.

Heald warp stop motion – In the event that a warp thread is broken, the corresponding heald wire falls, making contact with an electrode set on the bottom lath of the harness. This closes an electrical circuit, causing the weaving machine to stop

Mechanical warp stop motion – In the event of thread breakage, the dropper falls into a gap between the teeth of two dropper rods (one of which is stationary; the other, reciprocating). The movable lock-out rod initiates a signal to stop the weaving machine.

Safety devices – These are provided on weaving machines primarily to prevent defects in the fabric (e.g. warp stop motions, wefts forks, weft feelers, and also devices to detect breakage or misoperation of moving parts).

Safety devices against the breakage of warp threads – These are used over the length of the picking device to prevent weft thread breakage caused by the sley to the fabric fell when in its beating-up position., This avoids *unproductive weaving machine time and defects in the fabric.* It is to be found in various devices: collapsible sleys and the key mechanism on shuttle weaving machines, picking device detectors in a receiving box following weft flight through the shed on shuttleless weaving machines.

Warp protector in automatic shuttle weaving machines – This is used to prevent weft thread breakage: the warp protector dissipates the kinetic energy of the sley by striking shock-absorber links. The shuttle stops before it reaches the fabric fell, thereby avoid damage to the warp threads. Other safety devices combine elements of a folding reed and a warp protector.

Warp stop motion – This checks the integrity of each warp thread and stops the weaving machine in the event of warp thread breakage. There are several types of warp stop motions: dropper and heald, mechanical and opto-electric.

Weft feelers – Using weft forks, these detect the weft moving through the shed, preventing 'short weft in the fabric'. Weft feelers are positioned at the temple, at the exit of weft from the shed. A sprung feeler presses against the fabric fell. In the event of a missing weft in the fabric, the feeler is free to move between warp threads, which closes electrical contacts and sends a signal to stop the weaving machine.

Weft forks of electrical, high-frequency action – On water-jet weaving machines, a damp thread, bridging electrodes, closes an electric circuit. In the absence of a thread (an open circuit), the weaving machine stops.

Weft forks (electromechanical) – In a lateral weft fork, a sprung feeler is pressed by the weft thread. In the absence of weft pressure, the feeler is free to turn, closing electric contacts and stopping the weaving machine.

Weft forks (photoelectric) – A beam of light is directed at a reflector on the head of the rapier. In the presence of a weft thread, a lever covers the reflector. In the absence of a weft thread, a spring turns the lever on to a stop block, exposing the reflector. The beam of light hits a photodetector, and the resulting electrical signal stops the weaving machine.

Weft forks on shuttle weaving machines – There are two variants: located either at the edge of the shed, or inside the shed. *Lateral* weft forks (placed on one side of the warp) can detect the presence of weft in the shed following shuttle insertion into the shed after *two* picking operations, which exposes the fabric to 'shed with missing weft'. Weft forks located *centrally* detect *every* weft inserted in the shed.

Weft forks on shuttleless weaving machines – These can be located in several places: at the insertion side of the shed, in the shed, or at the exit from the shed; they detect every weft inserted in the shed.

Weft forks on the weaving machine – These detect the presence of weft in the shed and signal the weaving machine to stop in the event of weft breakage, or signal an automatic pirn change device. Structurally, weft forks differ according to the principle of their action: mechanical, electromechanical, opto-electric, piezoelectric, triboelectric, high-frequency etc.

Chapter 9

Direct drive of the main shaft of the weaving machine – This consists of an electric motor and two drive gear wheels ('pinion' and 'wheel'). The *direct drive* of the main shaft of a weaving machine switches on the electric motor at start-up and switches it off at the end of a run; it drives the main shaft of the weaving machine by means of a frictional clutch with a *constantly running electric motor*.

Electric drive of a weaving machine – This consists of an electric motor, a coupling or other system of direct drive to the main shaft of the weaving machine, a starting system and a brake.

Indirect drive of the main shaft of a weaving machine – On automatic weaving machines, the electric motor rotates a friction disk, by means of a gear wheel, and a thrust disk (which together form a clutch) and also rotates the main shaft. These can be of various constructions.

Weaving machine brakes – Various types of brake are used to stop a weaving machine quickly: electromagnetic, frictional shoe or band, disk, or cone.

Chapter 10

Coefficient of productive time in weaving machine operation – This is given by the ratio of time of actual operation of the weaving machine to all estimated time. The magnitude of this coefficient depends on the idle time of the weaving machine. Idle time is categorized as technical (repair, etc.) and technological (breakage of threads and warp beam installation, etc.).

Co-ordination of operation of the mechanisms of the weaving machine – It is convenient to indicate the co-ordination of the operational phases of the basic working mechanisms of a weaving machine according to the cycle from 360° or 720° of the angle of rotation of the main shaft of the weaving machine.

Minimum primary setting tension of warp threads – This level of warp tension is indicated by the absence of any noticeable sagging of droppers and false stoppages of the weaving machine (and also the evidence of obtaining a clean shed).

Normalization of the weaving process – This defines, analytically and experimentally, the optimum parameters of all elements of the elastic system of fabric formation (ESFF) and the rational parameters of adjustment of the basic working mechanisms. The primary parameters set up on a weaving machine (installation) are decided on the basis of the type of weaving machine and are such that will produce a good quality structure of the desired fabric in the most productive manner.

Productivity of a weaving machine – This is defined by the quantity of fabric made by the weaving machine for the time unit, taking into account idle time, m/h, or expressed in thousands of picks/h, or productivity expressed in square metres, m^2/h, or can be given by the consumption of meters of weft/h.

Technique of normalization of weaving process – In the *first stage*, this refers to the optimum primary setting of parameters calculated for the given type of weaving machine, for the development of the required structure of the fabric from a given yarn. In the *second stage*, this refers to the calculation of the primary setting parameters of ESFF geometry relative to the neutral line (NL) on the weaving machine. In the *third stage*, this refers to the objective control of non-uniform insertion of warp threads in the fabric and is resolved by experimentation. In the *fourth stage*, this refers to the evaluation of the magnitudes of optimum primary setting parameters of the weaving machine and is resolved experimentally with the use of tensiometric equipment.

Warp tension measuring method – *Electronic equipment* is used to obtain a visual representation of the dynamic variation of warp thread tension (tension oscillograms) in the ESFF during the weaving of the fabric.

Chapter 11

Control-lines (CL) machines – These machines operate lines that carry out the technological processing of the fabrics and fabric transport functions. There are three types of CL: continuous, periodic action and three-branch. A CL consists of a store for fabric rolls, unwinding devices, the sewing-machine, two or three control-registration tables, interoperational fabric compensators, measuring/packing machines or winding devices. CL machines also maintain a record of the rating and quantity of the fabric, clear the ends of loose threads, and removes oil stains by means of washing-up liquid and an aspirator pistol-spray. A CL machine rectifies all production defects, regardless of their nature, and has absolute control over the fabric produced.

Control of the weaving process – This involves the use of gauges to control the speed of operation of the equipment, the extent of short-term idle times due to warp and weft of breakages, machine faults, replenishment of yarn packages, machine idle time (say, for the cleaning of machines, etc.), and the quantities of production achieved.

Control of the quality and quantity of the fabric – After production on weaving machines, the grey (unfinished) fabric is sent to the control department. Here, the quality of the fabric is assessed (either manually or mechanically), and the fabric is prepared and packed ready for despatch either to the finishing factory or to the finished goods warehouse.

Defects of grey fabric

Bad selvedges – Selvedges that are wavy, non-uniform, flabby or hard occur as a result of weak or non-uniform thread tension of at the selvedge; incorrectly chosen interlacing or drawing-in of threads in the heald eyes, reed or droppers; a faulty selvedge mechanism in shuttleless weaving machines.

Pairing or *grouping* (reediness) – Warp threads are grouped in bunches; the reason for pairing or grouping can be the wrong choice of reed, the improper drawing-in of threads in the reed dents, or an insufficient degree of shed levelling (insufficient imbalance).

Drag-in – At the change of a pirn, the end of the previous pick is drawn into the shed and becomes woven into the fabric. The cause of this is incorrect adjustment of the temple scissors.

Drawing-in failures – Broken interlacing along fabric appears as a result of a faulty shedding device or on start-up of the weaving machine without a broken warp thread having been repaired.

Longitudinal stripes – This is when warp threads do not interlace correctly with weft threads over some length of the fabric, or when floats of different length are formed. This occurs when the tension of a group of warp threads is weakened following removal on breakage, or when they have been incorrectly drawn into the reed.

Missing warp ends – This is when one or several warp threads are missing in a section of the fabric and is caused by a fault in a dropper-type warp stop motion device.

Reed mark – This is a rarefied (widened) strip along the fabric caused by the curvature of a bent reed wire.

Ring temple punctures – Lines are formed due to punctures along the fabric edge made by a faulty needle temple.

Short weft – Incomplete weft along the width of fabric and *lacing* are characteristic of the weft falling short in the shed; also, the end of the weft may bend back. Possible reasons are faults in the weft feeder, or excessive or insufficient air pressure.

Starting marks (*start-up marks* or *set marks*) – These stripes extend across the whole width of the fabric and are made by reduced or increased weft density. They may occur in fabrics of low and average density when the weaving machine is re-started following a stoppage.

Unequal picking – This causes full-width stripes of varied weft density. A principal cause is a fault in the warp or cloth regulator.

Warp floats – These are interlacing defects in a small section of the fabric and occur because a group of warp threads float over one or several weft threads (e.g. when the end of a broken warp thread braids with adjacent threads). The reason for this is a fault in a dropper-type warp stop motion device.

Weft floats – These are weft threads over a small area of the fabric which do not interlace with the warp and appear in the form of loops on the face or the back of the fabric. These can occur due to a dirty shed, the incorrect timing of weft insertion in the shed, sagging warp threads, or incorrect positioning of the temples;

Weft loops – These loops are protruding and (more often at selvedges) unstraightened or twisted weft threads. They are caused by excessive twisting, insufficient humidity, or insufficient weft tension.

Weft stripes of higher density – These are stripes across the width of a fabric that have a higher weft density. This is due to faulty warp or cloth regulators.

Weft stripes of lower density – These are stripes across the width of a fabric that have a lower weft density. Ths is due to faulty warp or cloth regulators.

Equipment for the control of woven fabric quality – On the *defect-registration* machines, the fabric is removed from the cloth roller by means of system of cylinders on a viewing table. The viewing table has a glass base with a back-lighting device. After inspection, the fabric is turned around a directing cylinder and remains within the stacker for collection on the table.

Sorting of the grey fabric – This is based on defects of appearance and on indicators of physico-mechanical properties. A score is given for a sample length of fabric according to the total number of defects counted on it. Fabric control is carried out by defect-registration machines.

Chapter 12

Cart for the entering (drawing-in) of the warp – This is used to transport warp beams, together with the devices for drawing-in the thread (harness, reed, droppers).

Cart for woven fabric rolls – This is used for moving fabric rolls. The carts are provided with horizontal hooks and are moved by electrotractors.

Carts for warp beams – Warp beams are transported on floor carts by means of an electrotractor. Carts can be provided with basic rollers or with a hydraulic lift (as the warp beams can be heavy).

Product transportation in weaving manufacture – Threads for weaving manufacture come in a variety of package types depending on the raw material and configuration (spinning pirns, bobbins, skeins and others), and are delivered by road or rail. Electrotractors should not exceed 550 mm in width, and the speed of their movement in processing shops should not exceed walking speed.

Spike carts for bobbins – These are special carts with spikes ('fir-trees') that carry between 40 and 100 bobbins. The carts can be moved by electrotractors or floor chain, or suspended conveyor means, or by a magnetic transport system set in the floor.

Index

actual productivity, 166
air jets, 100–2
  features of weft insertion, 102–3
  weft insertion, 101
ATPR weaving machine, 139
automated control systems of weaving manufacture (ACSWM), 171
automatic warp brakes, 21

bad selvedges, 173–4
battery pirn supply, 67
battery type weft supply, 67–9
  weft pirns, 68
beat-up peak, 158
beating-up, 109
Beridot dropper warp stop motion, 135, 144
B.I. Damaskin mechanism, 70
bottom half-shed, 41

circular weaving machine, 88
closed shedding, 43
coefficient of useful time, 167
condensing method, 109
continuous formation, 2–3
continuous weft insertion, 83
control lines (CL) machines, 177
conventional thread length, 13–14
couples (reediness), 173
cyclogram, 46–7
  shedding cycle, 47

dobbies, 54–7, 62
  open-shed rotary dobby, 57
  single-lift dobby, 55
  staubli high-speed knife open-shed double-lift dobby, 56
DR-2 (Russian) machines, 175

Draper weaving machine, 131
drawing-in faults, 173
dwelling phase, 43, 44

elastic system of fabric formation (ESFF), 5–15, 18, 24, 25, 28, 32, 152
  individualisation, 31
  stiffness coefficient of a single warp thread, 6
  stretch deformation of threads in let-off zone of the warp beam, 9–13
  thread and fabric length in the elastic system of fabric formation, 13–15
  warp thread tension, 8
electric feelers, 67
electromagnetic drive, 106
electrotractor, 181
Elitex, 22, 23, 111
embedding method, 109
equivalent thread length, 13–14
Euler's law, 10

fabric cell, 114
fabric formation, 2
  zone, 116
fabric sorting, 175
finished fabrics
  movement in weaving manufacture, 179–81
  means of transportation of the yarn and fabric, 180
  product transportation, 179
  raw materials and outputs transportation, 179–81
frictional warp regulator, 31
'full-width' method, 109

205

Index

gauges, 65–7
  weft feeler, 66
grey fabric
  defects and quality assurance
    defects, 172–4
    quality assurance, 174–5

Hayashi dropperless contactless warp stop motion, 137
hollow pirn, 71–2
Hunt regulator, 30

idle time, 167
inertial method, 106
Inertial Small Picking Device (ISPD), 92–3, 95
International Textile Service (ITS), 174

Jacquard machine, 58–61, 62
  configuration, 58
  double-lift open-shed, 60
Jentilini-Ripamonti machine, 112

Keighley type mechanism, 128
Kovo pneumatic weaving machines, 130, 131

leasing device, 48
length of weft threads, 167
levelling size, 154
longitudinal stripes, 173

machine-yarn-fabric path (MYFP), 153
mechanical feelers, 67
microshuttle, 89–92
  drive, 90
  drive variants, 92
  multiphase weaving machine elements, 91
  weft on multiphase weaving machine picking, 91
missing warp ends, 173
missing weft, 65
monorail electric drive, 181
multicolour (multiweft) weaving machines, 3
multiphase weaving machine, 3
multishuttle mechanism, 72, 74–6

revolving shuttle magazine, 75
unilateral four-shuttle mechanism, 74

Neumann's Textima weaving machine, 94
neutral line (NL), 41, 47
Northrop mechanism, 69
Northrop weaving machines, 135
Novostav weaving machine, 94, 149

open shedding, 41–2
oscillogram analysis, 157–65

passive shuttle *see* projectile
periodic formation, 2
periodic weft insertion, 83
picking, 3
  shuttle weaving machines, 85–7
  schematic diagram, 86
pirns
  changing mechanism, 69–71
  schematic diagram, 70
  safety devices of automatic change, 71
planetary directing mechanism, 105
pneumatic (air-jet) weaving machine, 3, 22
pneumatic-rapier weaving machines, 3
pneumatic-rapier weft insertion, 103–5
  method of picking, 104
point beat-up, 2
projectile, 92–5
  electromagnetic picking device, 95
  picking device of a weaving machine, 93
  weaving machines, 3
  weft insertion on the Textima weaving machine, 94

rapiers, 95–100
  drive designs, 98
  method of continuous of weft insertion, 99
  picking methods, 96
  weaving machines, 3
raw materials
  movement in weaving manufacture, 179–81

means of transportation of the yarn and fabric, 180
product transportation, 179
raw materials and outputs transportation, 179–81
reed marks, 173
ring temple punctures, 173
Roper regulator, 26
Ruti scissors, 71

safety devices
  comparative analysis, 144
  devices for prevention of warp thread breakage, 142–4
    controller of weft picking device, 143
    warp protector of shuttle weaving machine, 143
  warp stop motions, 136
  weaving machines, 135–44
  weft controllers, 137–42
    central weft forks of shuttle weaving machine, 138
    weft feelers, 141
    weft forks on shuttleless weaving machine, 140
Schönherr-type negative mechanism, 130
shed closing phase, 44
shed opening phase, 44
shed with missing wefts, 138
shedding device, 3, 49
  comparative analysis, 61–2
shuttle, 72
  continuous weft insertion, 88–9
    circular weaving machine, 88
  positive drive
    ribbon weaving machine, 87
  replacement mechanism, 73
shuttle weaving machines, 3, 135
shuttleless weaving machines, 3, 76–8, 135
  weft changing mechanism, 78
Sick-Scan-System Ko-Re-Tra (Germany), 177
single-colour (monoweft) weaving machines, 3
single phase weaving machine, 2

sliding feelers, 66
start-up marks, 174
'Stema' product lines, 177
stiffness coefficient, 6–7
Sulzer weaving machine, 93, 131, 139, 143
Swiss Textile Institute of Zurich, 174

tape weft battery, 68
tappet pack setting, 53
tappet shedding, 49–54, 61
  construction of the sections of a tappet, 52
  independent movement of the heald shafts with a positive tappet drive, 51
  independent movement of the heald shafts with a tappet gear, 51
  tappet gear, 50
  tappet set, 53
Taylor and Buckley system dropper, 135
technical idle time, 167
tension measuring method, 157
tension oscillogram, 158
Textima weaving machines, 146
top half-shed, 41
torsion bar drive, 93
triboelectric weft stops, 140

unbalanced shed tension, 162
unequal picking, 173
'unifil' winding heads, 69

variable tension, 122

warp brakes, 20–4
  automatic brake for a warp beam, 22
  friction brakes, 20
warp delivery, 18
warp feeding, 18
warp floats, 173
warp regulators, 24–34
  belt type variator with continuous warp release, 29
  frictional warp regulator, 33
  hydraulic warp regulator, 31
  planetary warp regulator, 27
  worm warp regulator, 25

Index

warp release
  brakes and regulators comparative analysis, 36–7
  equilibrium of a moving back-rest, 34–5
    action of forces of tension of warp threads, 35
  overview, 17–20
    forces operating on warp beam., 19
  warp brakes, 20–4
  warp regulators, 24–34
  warp threads stabilisation and mode of release, 35–6
  weaving machine mechanism, 17–37
warp shedding, 40–62
  device classification, 49
  device comparative analysis, 61–2
  dobbies, 54–7
  Jacquard machine, 58–61
  parameters of shed, 40–4
    phases of warp threads, 44
    schematic diagram, 42
  tappet shedding, 49–54
  warp threads elongation, 44–8
warp tension
  brakes and regulators comparative analysis, 36–7
  equilibrium of a moving back-rest, 34–5
  overview, 17–20
    forces operating on warp beam., 19
  warp brakes, 20–4
  warp regulators, 24–34
  warp threads stabilisation and mode of release, 35–6
  weaving machine mechanism, 17–37
warp thread tension, 164
warp threads
  elongation in shedding, 44–8
    thread tension variation, 46
  stabilisation and mode of release, 35–6
  stretch deformation in let-off zone of the warp beam, 9–13
    warp let-off zone, 11
water jets, 102
  features of weft insertion, 102–3
  weaving machines, 3, 140

weft insertion, 101
weaving
  warp shedding, 40–62
    device classification, 49
    device comparative analysis, 61–2
    dobbies, 54–7
    Jacquard machine, 58–61
    parameters, 40–4
    tappet shedding, 49–54
    warp threads elongation, 44–8
weaving machine, 1–15
  advantages and disadvantages, 15
  apparatus for setting up the parameters of MYFP, 156
  classifications, 2–4
  comparing operating conditions for natural and synthetic fibres, 168–9
  coordination of weaving cycles, 165–6
  elastic system of fabric formation (ESFF), 5–15
  estimating drawing-in parameters, 153–5
  evaluating warp thread tension by oscillogram analysis, 157–65
    compact plain-weave fabric of higher strain, 161
    dense fabric of plain weave of average strain, 159
    extreme unbalanced tension oscillograms of warp threads, 163
    idealised models of tension variation of warp threads, 158
  factors affecting the productivity, 166–8
  glossary of terms applied to, 185–204
  mechanisms, 4–5
    machine–yarn–fabric path (MYFP), 4
  normalisation process for weaving operations, 151–3
  parameters for specific woven fabric structures, 151–69
  safety devices, 135–44
    comparative analysis, 144
    devices for prevention of warp thread breakage, 142–4

Index

warp stop motions, 136
weft controllers, 137–42
setting up parameters for the machine-yarn-fabric path, 155–6
supply of weft, 64–82
  battery type weft supply, 67–9
  changing pirns mechanism, 69–71
  gauges, 65–7
  hollow pirn change, 71–2
  multishuttle mechanisms, 72, 74–6
  overview, 64–5
  safety devices of automatic pirn change, 71
  shuttle change, 72
  shuttleless weaving machines, 76–8
  weft tension measuring device, 78–81
verification of parameters for the ESFF and MYFP, 156–7
non-uniform crimp of warp threads in the woven fabric, 157
warp release and tension control, 17–37
  brakes and regulators comparative analysis, 36–7
  equilibrium of a moving back-rest, 34–5
  overview, 17–20
  warp brakes, 20–4
  warp regulators, 24–34
  warp threads stabilisation and mode of release, 35–6
weaving machine brakes, 148–9
  illustration, 148
weaving machine drives, 145–6
  combined start-up and braking mechanisms, 149
  comparative analysis of different loom drives, 149
  mechanisms and types, 145–50
  mechanisms for driving the main shaft of weaving machine, 146–7
  illustration, 147
  weaving machine brakes, 148–9
  illustration, 148
weaving manufacture
  raw materials and finished fabrics movement, 179–81

means of transportation of the yarn and fabric, 180
product transportation, 179
raw materials and outputs transportation, 179–81
weaving technology
  further reading on, 183–4
  glossary of terms applied to, 185–204
weft
  battery type weft supply, 67–9
  changing pirns mechanism, 69–71
  gauges, 65–7
  hollow pirn change, 71–2
  multishuttle mechanisms, 72, 74–6
  overview, 64–5
  safety devices of automatic pirn change, 71
  shuttle change, 72
  shuttleless weaving machines, 76–8
  supply on weaving machine, 64–82
  weft tension measuring device, 78–81
weft controllers, 137–42
weft feelers, 65
weft floats, 173
weft forks, 139
weft insertion, 3, 83–107
  air and water jets, 100–3
  different method comparison, 106–7
  electromagnetic drive, 106
  inertial method, 106
  methods, 83–7
    shuttle motion on the weaving machine, 84
  microshuttle, 89–92
  pneumatic-rapier, 103–5
  projectile, 92–5
  rapiers, 95–100
  shuttle, 88–9
weft loops, 173
weft stripes, 118
  higher density, 173
  reduced density, 173
weft tension
  measuring device, 78–81
    weft measuring mechanisms, 80
    weft tension compensators, 79

woven fabric, 1
    quality control, 171–7
        defect registration machines and shearer-cleaner machine, 176
        defects in grey fabric, 172–4
        equipment, 175–7
        grey fabric quality assurance, 174–5
    weaving machine parameters, 151–69
    width, 3
woven fabric formation
    comparative analysis of the methods of fabric forming, 123
    fabric-forming mechanisms, 112–14
        fabric-forming elements, 113
    formation of woven fabric cell, 114–16
        'full-width' beating-up of weft, 115
    methods of easing of fabric formation, 122–3
    parameters, 116–21
        microsection of woven fabric along warp threads under variable tension of shed branches, 117
    principles and methods, 109–24

crank mechanism for reed drive, 111
tappet type sley mechanisms of shuttleless weaving machines, 112
ring temples, 121–2
    schematic diagram, 122
woven fabric take-up, 125–33
    comparative analysis of methods, 132–3
    mechanisms, 125–31, 132
        fabric take-up mechanisms with different actions, 128
        mechanisms for winding of woven fabrics, 132
        weft distribution in woven fabric, 126
        winding woven fabric on cloth beam, 131–2

yarn, 3
    reduction methods, 48

zone of formation (ZF), 117, 118
zone of stable wefts (ZS), 118

CPSIA information can be obtained at www.ICGtesting.com
Printed in the USA
LVOW07*1533120913

352190LV00038B/217/P